Like ghostly sharks
hunting by twilight,
the long-nosed Spitfires
winged eastward,
ready to flare their deadly flames,
to seek and destroy the enemy
wherever he appeared.

FLY FOR YOUR LIFE

The incredible story of Wing Commander
Robert Stanford Tuck,
immortal air warrior, hero and survivor
of the Battle of Britain.

A SELECTION OF
THE MILITARY BOOK CLUB

THE BANTAM WAR BOOK SERIES

This series of books is about a world on fire.

The carefully chosen volumes in the Bantam War Book Series cover the full dramatic sweep of World War II. Many are eyewitness accounts by the men who fought in a global conflict as the world's future hung in the balance. Fighter pilots, tank commanders and infantry captains, among others, recount exploits of individual courage. They present vivid portraits of brave men, true stories of gallantry, moving sagas of survival and stark tragedies of untimely death.

In 1933 Nazi Germany marched to become an empire that was to last a thousand years. In only twelve years that empire was destroyed, and ever since, the country has been bisected by her conquerors. Italy relinquished her colonial lands, as did Japan. These were the losers. The winners also lost the empires they had so painfully seized over the centuries. And one, Russia, lost over twenty million dead.

Those wartime 1940s were a simple, even a hopeful time. Hats came in only two colors, white and black, and after an initial battering the Allied nations started on a long and laborious march toward victory. It was a time when sane men believed the world would evolve into a decent place, but, as with all futures, there was no one then who could really forecast the world that we know now.

There are many ways to think about war. It has always been hard to understand the motivations and braveries of Axis soldiers fighting to enslave and dominate their neighbors. Yet it is impossible to know the hammer without the anvil, and to comprehend ourselves we must know the people we once fought against.

Through these books we can discover what it was like to take part in the war that was a final experience for nearly fifty million human beings. In so doing we may discover the strength to make a world as good as the one contained in those dreams and aspirations once believed by heroic men. We must understand our past as an honor to those dead who can no longer choose. They exchanged their lives in a hope for this future that we now inhabit. Though the fight took place many years ago, each of us remains as a living part of it.

FLY FOR YOUR LIFE

The Story
of
R.R. Stanford Tuck, D.S.O.,
D.F.C., and Two Bars

by Larry Forrester

BANTAM BOOKS
TORONTO • NEW YORK • LONDON • SYDNEY • AUCKLAND

*This low-priced Bantam Book
has been completely reset in a type face
designed for easy reading, and was printed
from new plates.*

FLY FOR YOUR LIFE
A Bantam Book

PRINTING HISTORY
*First published in Great Britain in 1956
by Frederick Muller Ltd.*

*Bantam edition / March 1978
2nd printing ...... March 1978
3rd printing .... December 1981*

Map by Shelley Drowns.
Drawings by Tom Beecham.

*All rights reserved.
Copyright © 1956, 1973 by Larry Forrester.
Foreword copyright © 1973 by Sir Max Aitken.
Illustrations copyright © 1978 by Bantam Books, Inc.
This book may not be reproduced in whole or in part, by
mimeograph or any other means, without permission.
For information address: Bantam Books, Inc.*

ISBN 0-553-20391-6

PRINTED IN THE UNITED STATES OF AMERICA

12 11 10 9 8 7 6 5

TO
MY WIFE
and
EIGHT OTHER PROUD TRAVELERS
OF THE INNER CIRCLE

FOREWORD
By Sir Max Aitken, Bart., D.S.O., D.F.C.

It may all seem very far away now, farther even in feeling than in years. It was certainly a much simpler world in 1939 and in the years immediately following.

Every problem was reduced to one problem—that of survival for the individual and the nation.

The story of Bob Stanford Tuck's war is one of the epics of that age and in the pages of Larry Forrester's narrative it lives again so vividly that the reader is inevitably caught up into the excitement.

What is more, the personality of the man comes over, the elegant swashbuckler I knew so well at the time, who was fanatically dedicated to the task at hand and who was, now and then, quite a headache to his service superiors.

Fly For Your Life recalls the exploits of a magnificent pilot, a legend in Fighter Command and famous far beyond it. It is a vivid, enthralling story.

Bob Stanford Tuck deserves it.

3rd November, 1972.

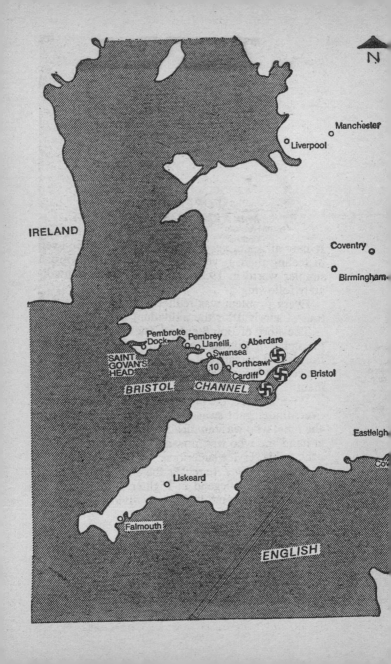

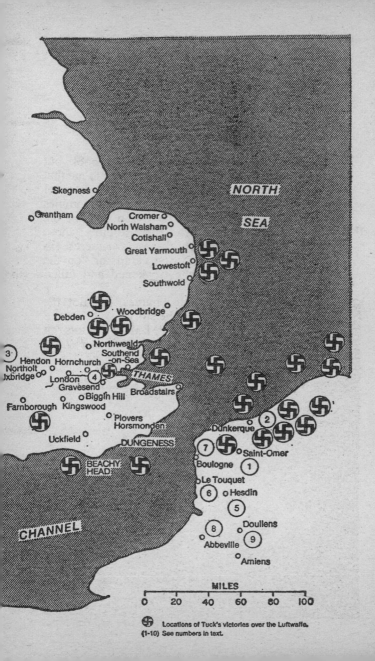

Locations of Tuck's victories over the Luftwaffe.
(1-10) See numbers in text.

There are no fictitious characters in this book, but there are a few fictitious names. It seems to me that so long after the war it would be needlessly cruel to re-awaken anguished memories for the families of those Royal Air Force men who did not die quickly or cleanly, or who died stupidly; those who contracted unpleasant disease or suffered extreme hardship in Nazi prison camps or "on the run"; the one or two who weakened and failed their comrades. . . .

So I have changed some names, but not the facts. The facts are part of the story.

LARRY FORRESTER
March 1, 1956

"They had that restless spirit of aggression, that passion to be at grips with the enemy, which is the hallmark of the very finest troops. Some—like 'Tin Legs' Bader, 'Sailor' Malan and Stanford Tuck—were so fiercely possessed of this demon, and of the skill to survive the danger into which it drew them, that their names were quickly added to the immortal company of Ball, Bishop, Mannock and McCudden."

—*official Royal Air Force History, Vol. 1,* *(The Fight At Odds) by Denis Richards.*

Wing Commander Robert Stanford Tuck,
D.S.O., D.F.C. and two bars

1

After lunch Blatchford and Tuck had one more beer together.

"Like old times," Tuck said as they raised their tankards.

"Sure," the Canadian agreed. But it wasn't. Now their talk was too quick, too fiercely gay, and their laughter loud. Each remembered much but kept it to himself—because there were so many names that could be mentioned only with a pang.

It was January 28th, 1942, a dank, gray day. Normal operations were impossible. The airfield at Biggin Hill was shrouded in drizzle and mist. A mean and narrow day, infinitely remote from those they had shared in that blazing, screaming summer of 1940. . . .

They had not seen each other for many weeks. "Cowboy" Blatchford now commanded a fighter wing at Digby, Lincolnshire, 180 miles away. With his squadrons grounded by the weather, he'd taken this chance to hedgehop across country and drop in for lunch.

He was a dumpy, toughly good-looking Albertan in his middle twenties, with a Westerner's rolling gait and measured drawl. As he talked and laughed, little wrinkles went in and out of his plump cheeks and his brown eyes seemed to flash in emphasis of a word or phrase. But today he was subdued.

"January," he said, slanting a look out of the window down the long, dark valley which bounded the station, "is just one long Monday morning. You've got a hangover from Christmas, and nothing to look forward to until spring—and that seems a *bloody* long way off!" He had a wry grin, and spoke without ardor, so his friend knew he was merely bored with enforced inactivity.

"Come with us this afternoon," Tuck suggested. "Do you the world of good, Cocky." He nodded across the room towards a young Canadian flying officer, tall as a candle, who was emerging from the dining-room. "I'm taking Harley. It's just right for a 'Rhubarb.' "

"Rhubarb" was the name given to a type of operation specially designed for cloudy conditions. Singly or in small groups, the Royal Air Force's fighters darted across the Channel, dived out of the overcast to shoot up selected targets, then climbed back into the murk before the Messerschmitts appeared. These activities, which demanded considerable skill and experience in navigation and marksmanship, had been standard "rainy day" procedure for more than eighteen months —part of Air Chief Marshal Sholto Douglas's policy of "leaning forward into France."

Intelligence kept station commanders well stocked with details and aerial photographs of likely "Rhubarb" targets—railway junctions, small factories, secondary harbors and the like. When the weather closed in, wing and squadron commanders picked the most convenient objectives and, with their most seasoned pilots, organized sneak attacks. They were seldom able to inflict heavy damage, but even in filthy conditions the Germans were given no respite, and were forced to strengthen their fighters, A.A. batteries and radar units in the west at the expense of their forces in eastern and southern Europe.

Today Tuck had selected as his target an alcohol distillery at Hesdin (5)*, some 21 miles inland from Le Touquet (6). Before lunch he and Harley had studied the "snapshots" and planned their approach from the French coast.

Cowboy wanted to join them, but felt he'd better get back to Digby, in case the weather closed in completely and left him stranded here. He came out to the aircraft to see them off, and just before they taxied away he climbed on to the wing of Tuck's Spitfire and stuck his

---

*Numbers in parentheses refer to locations on map.

head into the cockpit. His dark hair danced in the propeller's breath as he shouted above the deep, tuneless song of the engine: "Cheero, chum. Take it easy —keep it in your pants, huh?" This was his invariable leave-taking. His hand thumped Bob's shoulder once, then he jumped down. Tuck grinned and gave him a quick, casual wave as the two machines started lumbering through the grayness toward the take off point.

Wing Commanders Peter ("Cowboy") Blatchford and Roland Robert Stanford Tuck, were never to meet again. Within an hour Tuck's Spitfire lay smashed and black-smoking in a French field. And soon when spring came again, the Canadian's machine fell, doom-spinning, into the dark maw of the North Sea. . . .

Winging eastward, a hundred and fifty feet above the mist-veiled face of Kent, the long-nosed Spitfires were like ghostly sharks hunting by twilight. Under each tiny perspex dome the pilot's hands moved with custom-learned speed and lightness, checking various control settings in readiness for the coming action. Radiator to "warm;" gunsight illumination fully dimmed in the absence of sunlight; radio-transmitter set to "receive" with the volume well up, but the microphone inside the oxygen mask switched "off" so as to maintain silence until they reached their objective.

Lastly, the gun-button on the control column set to "fire." In dark, well-greased cells deep within the wings hundreds of 20-millimeter cannon shells and .303 machine gun cartridges lay sleeping; one touch on the button with the ball of the thumb, and in an instant they would awaken with a shuddering roar, hurtling ahead as almost solid streams of metal. Today these two fighters had been specially armed for their inflammable target: every second shell was an incendiary, and the bullets were of the type known as de Wilde, which combined the qualities of tracer and explosive with a solid armor-piercing core.

Thus, with much cumber and array, did two of Britain's air warriors go forth on that winter's day, at a time when the scales of fortune were beginning to tip in our favor after the long, hard siege. The Supermarine Spitfire fighter, designed by that audacious innovator R. J. Mitchell, and employed so skillfully and decisively in defending Britain against the great Luftwaffe swarms of 1940–1941, now had become a useful offensive weapon. And the most aggressive, tireless pilots who flew on the cross-Channel sweeps and "Rhubarbs" were those few survivors of the Battle of Britain who remained operational. Men like Bob Tuck.

At 25, he was both myth and mystery. He held the Distinguished Service Order, was the second officer in the Royal Air Force's history to gain a second bar to the Distinguished Flying Cross, and had more air victories than any other British pilot. (Air Ministry's official credit was 29 enemy aircraft destroyed; Tuck's private tally was 35. In both respects he was at this time just ahead of his close friend, the fabulous South African, A. G. "Sailor" Malan.)

Legend described Tuck variously. In one version he would be gay, reckless in young power—a boyish grin behind the gunsight. Another would portray him as a flying fanatic, stern as a Caliph and psychopathically bloodthirsty. In fact, he was neither of those things, but a little of each.

The great mystery about Tuck was how he had managed to stay alive, how he kept his aggressiveness simmering, his eye true and his hand steady after years of unrelenting physical and nervous strain. He had been shot down four times, slightly wounded only twice. He had survived two air collisions. He had bailed out, crash-landed, "dunked" in the Channel. On various occasions enemy fire had set his aircraft ablaze, shattered his windscreen, blown the throttle lever out of his hand, ripped the oxygen mask off his face. Once a bullet had struck his thigh, rattling the loose change in his pocket and stoving a 1921 penny into the shape of a spoonhead. (He still carries this souvenir among his daily supply of coins, sometimes toys with it during conversation.)

A long, straight, dead-white scar slanting down his right cheek—legacy of a pre-war accident—seemed only to emphasize his dramatic, if unEnglish, handsomeness: a fraction under six feet tall, whip thin, whip strong, dark and graceful and elegant as a matador. His hair was gleaming black, and apart from a slight puff at the front, very flat. It was brushed, hard, almost straight back from his high, sloping brow and this, together with the prominent cheekbones and the rather long, knifelike nose, gave his head a curious appearance of streamlining and high speed.

His eyes were rich brown, steady and yet quick, with the merest hint of an Asiatic tilt at the corners. At times they seemed to hold tiny golden flecks—the sort of granulation often found in animals' eyes. Underneath were deep, charcoal circles. His chin was narrow, with knots of muscle on either side. Under the tanned skin of his cheeks the blood pumped almost coral-red —limning the long scar in vivid relief. His lips were thin, but strong in outline, quick to smile and reveal perfect teeth. On the backs of his long, slender hands the sinews stood out thick and taut and white.

His laugh was a tenor shout of pure glee, and his voice was without a trace of either accent or affectation —about all you could say of it was that it was middle-pitched, "English and educated," with a fairly wide and colorful vocabulary at its disposal. He always spoke quickly, as though perpetually excited. At certain times, however—particularly when addressing a stranger of any importance—a slightly artificial intonation crept in, a sort of over-preciseness that seemed to indicate some secret, still unconquered complex.

His slim, black mustache was meticulously trimmed, and it was common knowledge that he had his uniforms —even his battledress—tailored to measure in Saville

Row.* In three years of war he had grown swiftly older, thinner, paler; yet all the quickness, exuberance and unreasonable optimism of youth remained.

On the ground he seemed just a restless, voluble character with a huge appetite and a zest for stag parties: but once in the air some strange, schizophrenic process transformed him into a grim, cool and crafty destroyer. He had his own methods. It was like women cooking—he didn't use rules, he just knew. Those who flew with him felt almost spellbound. He spoke only when necessary and, apart from a frank, savage glee in the moment of a kill, displayed no excitement, no sign of any human emotion. The voice that rapped out orders over the radio-transmitter was curt, flat, impersonal. At times, downright sinister.

His main faults were vanity and impatience. Fortunately, neither detracted from his abilities as a pilot and leader—on the contrary, they were important ingredients of his success.

A complex man, this Tuck. The sort of man you could love readily, fear instinctively, and never quite get to know.

*   *   *

Now, on this routine operation, the Canadian Harley, like so many other young pilots before him, was boyishly thrilled to fly under the radio call sign "Bobbie Two." Though the air was bumpy beneath the rain clouds he snuggled in fairly close beside his leader and, ceaselessly making rapid, delicate corrections with stick, rudder and throttle, managed to fly in neat and steady formation.

They streaked over the familiar pointer of Dungeness dead on track, and immediately went down to twenty feet over the sea so that they wouldn't be picked up by enemy radar scanners. Visibility was decreasing, so they eased a little farther apart and concentrated on staying out of the water: in these conditions it was tragically easy to misjudge height and fly smack into a heaving wave.

*The street of London's finest custom tailors.

The Channel was rough. A stiff wind was whipping spray off the tops of the breakers like snow off mountain peaks. In what seemed a matter of seconds the time-cracked cliffs of France advanced, solidifying and growing out of the gloom. They held their planes down on the water until the last moment, then pulled up sharply, grazing the brink. As they sped on inland a few feet above the patchwork of fields, roads, woods and villages, long golden spears of tracer rose half-heartedly close behind. Their sneak entry had been detected by a coastal battery; at this instant, they knew, the entire defensive system would be jarring awake to the danger.

After a few miles they picked up a main railway line, running due east. They followed this without further incident until Tuck, map-reading from a folded chart on his left knee, spotted their turning off point. A tight turn to the left across some large, unkempt meadows brought them towards a second railway line with a high embankment. Beyond this lay their target.

According to plan they roared towards the embankment so low that their passing left a fluttering wake in the tall grass. Harley heard a sharp click in his earphones and then that flat, clippped voice: "Okay, Bobbie Two! Come up to line abreast and stand by to pull up. Over." Swiftly the Canadian flicked his transmission switch and acknowledged, then advanced his throttle and drew level with his leader.

"Pulling up now!" rapped Tuck.

As if the embankment were a slipway, the two Spitfires flew up it and launched themselves steeply at the rumpled canopy of clouds some eight hundred feet above. As they leveled off close under the grayness a number of 20-millimeter guns opened up from rooftops and sites in the grounds of the factory, only a few hundred yards ahead. This they had expected—A.A. batteries for many miles around were always alerted the moment intruders were spotted crossing the coast. They were not perturbed: a Spit in a dive was a small and exceptionally fast target, particularly difficult to hit in dull or hazy weather.

"Down we go!" Weird, that voice of his. Even at

this juncture Harley couldn't help noting that it had no cadence, that each syllable was pitched on the same note. Down went the noses, and their right thumbs took the first, light pressure on the gun-buttons. With amazing slowness the rods of tracer drifted up to meet them, and then suddenly whisked past—above, below, and on either side. Now they could see the four alcohol vats, clumsily camouflaged and looking like small gasometers, standing well away from the factory's buildings in a large compound. It was all exactly as it had looked in the Intelligence photographs. Except that now the whole scene was sharply tilted, and rushing towards them at nearly 300 miles an hour. . . .

Each picked a vat, and Harley opened fire first. His Spitfire's wings became barbed with spikes of blue and orange flame, spewing cartridge and shell cases. Towards the end of his first, long burst he was hitting squarely. Tuck, the more experienced marksman, held his blast until he knew he was well within range. The instant before he pressed the button he dropped his eyes from the gunsight's glowing reticle and, as he had done hundreds of times before, checked his turn-and-bank indicator. The black needle quivered in the dead center, telling him he was not in a skid—therefore his sight reading would be true, and if he had lined up properly he couldn't miss.

Above the first thunder of his guns he heard Harley's triumphant shout: "There goes mine!" Out of the corner of his eye he saw a black pall rising lazily, then his own vat crumpled like melting wax, issuing clouds of smoke and dark flame. They were doing 320 m.p.h. as they cut through the smoke scant feet above the wreckage. Harley broke off to the right, Tuck went left.

As his nose came round Tuck saw a long, low building with narrow, barred windows passing to the right in front of him, like a railway train. He felt sure it was a barracks. Another quick glance at the turn-and-bank indicator, a minute adjustment in the amount of rudder to correct the ghost of a skid, then he pressed the button again. The long burst swept the length of the building. His fire must have gone in at several of the windows for he distinctly saw his shells exploding in the dark interior. Then he was up, into the cloud and out of the claws of the flak.

He relaxed in his seat, set course west and eased his throttle back to give Harley a chance to catch up. After about a minute he came down out of the cloud and almost at once the Canadian called: "Okay, Bobbie One, see you now dead ahead. Closing up."

When they were together again, both apparently undamaged, Tuck tried to pick up the main road he'd marked on his chart as their way out. But the weather was worsening quickly now, and they couldn't see the ground for more than a few hundred yards in any direction. It was quite impossible to tell one highway from another, so they simply picked one and followed it westwards.

Suddenly they came upon a heavy military truck, squirted at it, and sent it rolling into a ditch. They had plenty of ammo left, so when they saw a tall, steel pylon looming ahead they swung a few degrees off to the north so that they could follow the line of high tension cables. This could be fun—if you were able to hit the insulators, great multi-colored flashes went searing away for miles across country. But today they were unlucky and all they could accomplish was an occasional shower of sparks.

So engrossed were they in this game that they failed to notice that the line of pylons curved slightly, taking them further and further north. All at once they skimmed some foothills and found themselves flying at not more than two hundred feet across a gradually widening valley, the floor of which was crammed with railway sidings, engines and rolling stock. At the end of the valley they could make out the drab mass of a large town, and beyond that the black tracery of derricks and masts silhouetted against the sea. Tuck knew it could only be Boulogne (7)—they were five or six miles to the north of the track he had planned for the homeward flight.

The area was notorious for its heavy A.A. defense. His first instinct was to make a quick turn of 180 degrees, strike back inland and then find his way back to base over a quieter section of the coast. But even as he began to apply turning pressure on the stick and rudder he saw ahead of him a large, modern engine. It was stationary: irresistible.

What followed is best described in his own words:

"I thought 'in for a penny, in for a pound,' and called Harley up into line abreast. We dived on that engine together. I think we both scored hits, and the whole issue disappeared in a tremendous cloud of steam.

"As we went through the steam, just for a moment I lost sight of Harley, so I made a quick turn to get out of his way. As I came into the clear again at about seventy or eighty feet, I think everything in the Boulogne area opened up on me.

"I saw right away that Jerry had his guns at the foot of the hills on either side of me, and they were shooting more or less horizontally across the valley. I was caught in their cross fire, and at this low altitude with a forty-five degree bank on, they just couldn't miss.

"Several shots smacked into the belly. One shell came right up through the sump, through the cooling system, and everything stopped dead. She started to belch black smoke, glycol and all sorts of filth. The windscreen was covered in oil, so I had to slam back the canopy and stick my head outside.

"I still had quite a bit of speed on—about 250— so I could stay in the air for perhaps a minute. But I knew if I tried to pull up to bail out, they'd blast me clean out of the sky. My only chance was to stay very close to the deck, and try to find a space to put her down in. I turned up the valley and kept weaving her from side to side, but I couldn't see the smallest patch of open ground.

"When they kept on pouring shells at me, I had a terrific temptation to climb her before the speed fell too low, but I knew the moment I got on to the skyline they'd have me like a box of birds.

"At this point I happened to glance over my shoulder, and there was Harley following, right on my tail. It looked as if he was going to get the daylights shot out of him, too. I called him up and said: 'Get out, quick! I've had it." But when I looked again, several seconds later, the clot was still there."

When Tuck's voice came again, it was no longer flat and impersonal, but a full-blooded roar: "Get out, you stupid twit! That's a bloody order!" Harley was so startled that he had yanked the stick over and was streaking out to sea before he had time to collect his thoughts. And then, in the sudden shock of realizing what had happened in those last thirty or forty seconds —that the great Stanford Tuck was vanquished at last, and going down into a hopelessly cluttered area where there seemed no chance of a safe landing—the young Canadian could only wipe at his stinging eyes with the back of his glove and say over and over again: *"Christ! Oh my Ch-r-ist!"*

Meanwhile Tuck's machine was gradually losing its momentum, dropping its nose in forlorn defeat. There was a terrible finality about that movement. Sixty feet . . . fifty . . . forty. He stuck his head out as far as he could, peering ahead. Still no sign of a clear space. They were still shooting at him from all sides, and with the engine silent he had an unpleasant sensation of vulnerability.

At twenty feet he faced the facts, braced his lean body and told himself: "This time I'm really done for." Taut and clammy, he sat back, watching the tracer coming at him, waiting. And then, incongruously, he thought—"Dammit, I won't be able to phone Joyce tonight!"

Joyce was the girl he was determined to marry—or had been, until that shell came through the engine. Joyce was serene, cool and English. She spoke softly, smelled sweet, and moved like a melody. Nothing ruffled her, nothing changed her. She seemed to him a part of that normal, permanent way of life he had never known, but dreamed of more and more over the last few months.

Seconds left now. In fleeting, sad thoughts he caught again the theme of that endless, noisy, summer battle —a year and a half ago now, but living on in him, an unforgettable, indispensable part of his being: the June sun that burned them awake, its light too strong for their weary eyes; the heady scents of gas, hot metal, rubber and cordite; the harsh screaming of their frantically climbing engines as the wide, thundering rivers of bombers poured out of the east; the Spits and Hurris darting down, glinting like schools of tiny fish. . . .

Then it all dimmed away, and there was only the floor of this wintry French valley reaching up for him —blank death inches away. His airspeed indicator was unwinding remorselessly towards the tiny red line on the dial which marked the speed at which the machine would cease to fly, would stall and dash itself to pieces on the ground. Numbness oozed into his bones, his limbs seemed packed in ice. He had long since accustomed himself to the idea of violent death, given up all hope of surviving this war. "Well—here it comes, any second now. . . ." His brain froze on these words, repeating them over and over again, like a record player with the needle stuck.

And then suddenly, miraculously, a narrow sward of deepest green loomed ahead of the right. A field! A long, lovely French field! He swung the plane around; unexpectedly the ground fell away for about a hundred feet. He dived toward the greenness, building up precious speed again. A chance, a good chance now— all he had to do was set her down correctly, with the wheels retracted so that she wouldn't roll too far or tip over on her nose if the ground were soft or if she hit a rut or snag. . . .

But the sudden, gleeful dive had given him a bigger increase in speed than he'd bargained for. Nearly 140 on the clock—55 m.p.h. above the normal landing speed. He was going to put her down all the same!

And then something flashed close over his head— something that made a sinister, whooshing sound. He lifted his eyes from the grass and dead ahead, at the far end of the field, saw a multiple 20-millimeter mounted on the back of a truck. The muzzles were fully depressed and spitting flame. He was coming down slap in the middle of one of the A.A. batteries, and the swine were still shooting at him—at almost point blank range!

Fury hammered at his temples. They could see he had no engine power—why didn't they leave him alone,

give him this last chance to save his life? He shoved the stick forward, lined up his sight—and by sheer force of habit checked the turn-and-bank indicator. Then he fired his last, short burst.

The whooshing sound stopped. He was still doing 120 but he was running out of field, so immediately he took a deep breath, put the stick forward and set her down. The first heavy impact jerked him against his safety straps and his brow struck the gunsight.

When his senses returned, the Spit had come to rest on a slight rise. His nose was bleeding and his right eye was rapidly closing. He struggled out of his harness and clambered on to the wing. A few yards away the gun-truck lay shattered and smoking, the mangled bodies of its crew strewn around it. His shooting, as usual, had been deadly accurate.

From both sides of the field, gray figures were running towards him. There was no hope of avoiding them. "The bastards will lynch me now," he thought

bitterly. For once, he had forgotten to bring a revolver
—he usually carried one in the leg of his right flying
boot. Fumbling in the cockpit he crumpled his chart,
and tried to set it on fire with his cigarette lighter.
But the lighter wouldn't work.

Zonk! He wheeled round as a bullet went through
the fuselage close by his elbow. One of the approaching
Huns had seen his intention and opened up with an
automatic rifle.

They were closing in from several directions. Mallet-
faced, jack-booted troopers, yelling angrily. He caught
the flash of a bayonet here and there, looked in vain
for an officer to whom he might appeal for fair treatment.

It was too late to reach into his tunic pocket and
throw away the Iron Cross he carried as a sort of
amulet. It had been given to him over a year before by
a vanquished Luftwaffe pilot in an English hospital—
but, of course, *they* wouldn't like that. . . .

So—by this totally unexpected, simple misfortune
—ended the fighting career of a pilot whose phenomenal luck had become a byword in the Royal Air
Force, whose extraordinary skill and unfailing energy
had moved newspaper correspondents to write: "Stanford Tuck is unbeatable in the air. . . ." "the immortal
Tuck . . ." "Tuck, the man they can't keep down. . . ."

Unbeaten in the air he remained, for the immortal
Tuck had been kept down, at last, by shells which
were aimed and fired *from the ground.*

Slowly he raised his hands, leaning back against
the fuselage so that the oncoming Germans wouldn't
see immediately the row of swastikas stenciled there:
twenty-nine of them. As they grabbed his arms and
started to drag him towards the wreckage of the gun-
truck, he thought wearily what a long, hard road he
had come, only for this.

2

St. Dunstan's College, Catford, is a mellow red-bricked building, snug in ivy and attended by drowsy oaks and elms. It has neat flower beds, an easygoing stream and a wide, friendly gravel drive. It is the kind of school the most intractable pupil would find hard to hate.

Roland Robert Stanford Tuck, the energetic younger son of Captain Stanley Lewis Tuck, head of an export business, did not take well to the quiet, leisurely life of St. Dunstan's. Since it would not permit him to hate it, he retaliated, by vowing that he would never *like* it. This vow—like most others he has made—has been diligently kept. Although undoubtedly Tuck now ranks among the College's most famous pupils, he has never returned for the briefest visit. It should be recorded, however, that the amiable old place has never invited him back.

All the same, Bob Tuck is in St. Dunstan's debt. In the classroom he could never sit still long enough to learn much though he seemed to have a natural flair for languages—but in the gymnasium, on the playing fields, and on the rifle range, the College gave him much that was to prove of the highest value throughout his air force career.

When he left at the end of the 1932 term, he was a keen gymnast and amateur boxer, average at cricket and lacrosse. He was an excellent fencer, a prize winning swimmer and easily the best shot of the St. Dunstan's rifle shooting team which had won the Sherwood Trophy at Bisley that year.

Shooting—and guns—were his passion. As a child he had never played with a toy pistol, for his father had given him an old Army Colt revolver and a dilapidated lightweight shotgun—both minus their firing pins, of course. From this the boy had built up a formidable private armory, which included antique pistols, powerful air rifles, a variety of sporting guns and even a primitive musket. All these later acquisitions were in good working order. He had acquired a practical knowledge of ballistics, and since ten years of age had been making his own ammunition in a smell shed in the three-acre grounds of the family home, "The Lodge," a rambling old house in the suburb of Catford, on London's southeastern fringe.

Captain Tuck had served in the Great War with the Queen's Royal West Surrey Regiment, and had a soldier's respect for firearms. From the start he sternly drilled Robert, and his elder son Jack, in the handling and maintenance of their weapons. He taught them, too, the fundamentals of deflection shooting, and made them practice until the process of lightning calculation and coordination of eye and hand became automatic. Jack was only a fair shot, but Robert, remarkably keen-sighted, quick of reflex, and a natural judge of speed and distance, at the age of twelve seldom missed a rabbit on the run or a pheasant on the wing. He enjoyed sport much more than routine shooting at fixed targets on the range.

Though his efforts with St. Dunstan's rifle team brought him occasional medals and cups, it was his "free lance" shooting which at one period earned him his largest and most unusual collection of trophies. When he was thirteen he was sent to spend part of his summer holiday on the turkey farm owned by his Uncle Victor and Aunt Grace at Esher, Surrey. On his first day there he learned that the compounds were constantly besieged by cats, and that now and then one or two raiders got through the wire and wrought havoc among the young birds.

From a vantage point amid thick shrubbery, well away from the house, he sniped at the prowlers with a 410 shotgun, loaded with homemade cartridges. After about a week, Aunt Grace had occasion to enter the garden toolshed: to her great horror, she found some thirty cats' tails hung in rows on the walls, like scalps in a Red Indian's tent.

Robert was packed off home in disgrace, with the beginnings of a guilt complex which was to torment him secretly for many years. Ever since, he seems to have been making constant efforts to obtain forgiveness from the cat world by adopting strays and lavishing kindness on them.

There were three females in the house at Catford—strikingly contrasting in appearance and character. His mother, Ethel Clara Tuck, was tiny, soft-eyed, and soft-voiced, gentleness personified. Yet she had great passive strength and infinite patience, a sort of family faith that never wavered, and became most apparent in time of crisis. She seemed to provide continuity for the family story.

There was no malice in her toward any living being. A gentlewoman, she could not believe the world was as violent and sordid as even the most restrained newspapers averred, who could pity the most savage animal or reviled criminal and try to find excuses for his most evil acts. Her husband and her children invariably addressed her as "Ma-ma," using the Victorian emphasis on the second syllable.

Robert's sister Peggy—five years older than he, and eighteen months older than Jack—was a vivacious, fiercely independent girl. Even at sixteen she managed to remain aloof from her brothers.

The third female was Aksinia, a middle-aged Russian. She had tight-drawn black hair, a thin, deep-lined face, melancholy eyes and a husky voice. She had come to England, when Robert was a baby to work in a company formed by Captain Tuck to trade principally with eastern Europe. This venture had crashed—a blow from which the family took years to recover, and which deprived Robert of the opportunity to go to college—but Aksinia had stayed on with the family as a sort of housekeeper and governess.

She was a strange, complex woman—prone to dark, impenetrable moods and long, sad silences, yet often gay and voluble; sometimes strict and unrelenting, at others overflowing with sweetness and sympathy. She told romantic tales of the Old Russia, some historically factual, others based on folk lore—which she embellished unashamedly and wove into wild, at times even macabre, fantasies.

Robert, the youngest child, was Aksinia's favorite. Even before he started school she had him speaking odd words and phrases of Russian, and soon afterwards began to teach him a sentence every day. At fourteen he not only could converse freely in the language, but could write it too. This, Aksinia continually told the family, would ensure his eventual fame and fortune, despite his lack of other, more practical, academic qualifications.

She could not have guessed the day he would find this knowledge invaluable. . . .

For many weeks after leaving St. Dunstan's, the boy made no effort to find employment. At dinner table each evening his father rumbled admonitions and Jack—firmly established in the general office of a London millinery manufacturer—tried in vain to explain how exciting was the life of a young businessman in the crowded canyons of the City. If a job was absolutely essential, Robert was determined above all else that it should not be in an office, preferably not indoors at all.

"Good Lord," cried the exasperated father one night, "d'you want to be a damned ditch digger?"

Robert looked up from his plate and, perfectly serious, answered: "What's wrong with that? Seems a pretty good life to me—plenty of fresh air, and traveling from place to place. . . ."

The very next morning Captain Tuck called him into the study.

"Robert—what about the Merchant Navy?"

The boy's eyes gleamed. "You *know* I'd love that, Dad! I'd have suggested it long ago, but I never thought you'd let me go."

"Can't have you lazing about here all your life."

"What about Ma-ma?"

"I've spoken to her. She's a bit upset, of course, but

she agrees it's all for the best." He paused, sighed, and suddenly looked very tired. "She'd have liked to see you at the university—I would, too—but you know very well that's out of the question. Be a year or two yet before we could afford it. Anyway, I'm not sure you'd make a go of it."

"I'd rather go to sea!"

"All right, that's settled. I'll make inquiries right away."

And it came about, less than one month later, that Cadet R. R. S. Tuck, feeling hideously conspicuous in a uniform that hadn't yet lost its tailor shop stiffness, walked up a gangway at Falmouth, Cornwall, to the deck of the 15,000-ton, twin-screw steamship *Marconi*.

A group of deckhands eyed this pink-cheeked stripling with frank amusement while he stood shuffling his feet, waiting for someone to greet him. After a minute or so, one of the oily-vested toilers took pity on him, jerked a thumb toward a companionway and called: "Down-a thar, matey. First Mate's cabin you want." Bob nodded his thanks and, lugging his shiny new suitcase, clambered below.

The First Mate's name was Philip Furlong. He was a stocky, lusty man and he welcomed the new cadet with a vigorous handshake. He surveyed the boy's gawky frame and sharp features and chuckled: "Well, well—little Tommy Tuck, eh?" The nickname stuck for many years.

"You've picked a good ship, son. Good skipper, an' all. Make up your mind to work and learn, and you'll find this a damn good life."

Furlong took him to meet the *Marconi*'s master, Captain Edward Wilkinson, with his round, red face and close-cropped, sun-bleached hair, seemed almost benign. The ship was clean and white and full of exciting noises and smells. Bob's morale rose in a spiral of enthusiasm.

He realized how very hard he would have to work: officers cadets stood normal deck watches like everyone else in the crew, and in their off-duty time they had to study and receive instruction from the master and ship's officers in navigation, seamanship, geography, astronomy, engineering and a host of other subjects. No keener cadet ever set sail from an English port.

The *Marconi* was a refrigerator ship with accommodation for twelve passengers, owned by Lamport and Holt and operating mainly on the South American "meat run." Occasionally, however, she was loaned under charter to the Union Castle Line to bring home fruit from South African ports.

Now she was bound for Capetown, Port Elizabeth and on up the coast to Durban. She had hardly passed over the bar before Bob's eyes started to ache strangely. Then his stomach tightened and began to heave. For two-and-a-half days he was hopelessly, miserably seasick. Rendered unfit for duty, he was filled with a deep, searing shame.

"Furlie," the First Mate, tended and comforted him with surprising gentleness. When he was able to totter on deck Captain Wilkinson gave him a quick smile: "Good lad—now you're ready to start, eh? Put him to work, Mr. Furlong."

It was a very different Furlie who now took charge of him, and in ripe and chesty tones ordered him about the ship for the next few weeks, taking delight in finding him the dirtiest, most uncomfortable tasks. Bob found himself alternately—and at times, simultaneously—smeared in oil, smothered in flakes of rust, streaked with paint, bespattered with red lead, soaked in spray and sticky with salt. And every time he lifted his head, there stood his taskmaster, legs astride and hands on hips, glowering down at him and ordering him to do the job all over again. At times his temper boiled and bitter tears welled up, but always he clenched his teeth and labored on.

Days and weeks merged. Four hours on and eight hours off. Muscles trembling with fatigue. Sleep—work—sleep—work. Shadeless days, and nights lit by thickets of stars. The soft hush-hush of sliding water as the ship cut deeper into the tropics. Seabirds' cries one dawning—and then, suddenly, there was Table Mountain dead ahead, crowned with fibrils of mist, and the town nestling beneath, bright and delicate as a toy. Even as he stared the First Mate's bulk materialized beside him, and he hurriedly got back to work. But now Furlie was smiling, the way he'd smiled on the day Bob first reported to him.

"Well—how's it look to you, Tommy?"

"Fine. Just fine, sir." He was surprised at the conviction he'd managed to get into his voice—in fact to his weary eyes it all looked unreal as a poster in a travel bureau's window. And then for no apparent reason Furlie dealt him a resounding smack on the shoulder and bellowed with laughter.

"You'll do, lad! You'll do!" cried the Mate. "You're skinny as a Scotch hen, but you're tough, boy, you're really tough!"

It was perfectly true. Bob was very thin, but he had the hard-bitten gameness and tireless strength of a sled-dog. He had accepted shipboard routine and discipline without complaint and shown that he didn't need to be given an order twice. He had won the Mate's respect—won a new and powerful friend. From now on life was to be easier.

As they went up the coast, sharks began to follow the ship. Bob was painting the rail one day when Captain Wilkinson came on deck, saw the dorsal fins close astern and shuddered violently. He didn't seem to notice the boy, standing only a few feet away.

"Brutes!" he snarled with astonishing venom. "Murderous, bloody brutes!" Then he turned on his heel and strode to the bridge.

It was Furlie who explained the cause of the skipper's revulsion.

"Proper mania with him—my *God,* how he hates 'em! And with reason, lad. Saw his best friend taken by a shark on the Brazilian coast. Good few years ago, but he's always dreamin' it over again—hearing his friend screaming, and seeing blood spreading on the water."

Bob was silent for a moment, following the curving path of one of the evil black triangles. Then he asked: "Is there a rifle on board?"

"Yes, of course."

"If I could borrow it I'm sure I could kill a few of these."

Furlie studied him, then nodded toward the bridge.

"I'm not giving it to you. Go and ask the skipper."

Somewhat uncertainly, Bob climbed to the bridge and made the request. Captain Wilkinson shook his bristly head.

"Waste of ammunition, m'boy. You could blaze away all day and never sink one of the brutes. Hides like steel, y'know. If your bullet strikes at the wrong angle it bounces off."

"I'd like to have a try, sir. I'm *sure* I could knock off two or three."

The captain's pale blue eyes lifted over the lad's shoulder, and cold hate flared in them as they rested on the flashing fins alongside the creamy wake.

"All right, Tommy. Mr. Furlong will give you a gun and twelve rounds. See how you do with that."

It was an old Army Lee Enfield .303. With the twelve rounds Bob sank five man-eaters—and was shocked and sickened by the carnage which ensued each time his bullet found its mark. The other sharks, maddened by the scent of blood, fell upon the dying fish and tore it to pieces in a matter of seconds.

Came a roar from the bridge: "First class shooting, that! Mister Mate—give him all the ammunition he wants!"

From then on, whenever the *Marconi* was in tropical waters, Bob's chief duty was shark killing. He became so skillful that members of the crew gathered daily to watch.

Using the exposed fin as a point of reference, he learned to judge the position of the pointed head under the water and, laying on deflection to allow for the distance which the target would travel before the bullet reached it, could kill nine times out of ten at up to sixty yards range. For this service—"cleaning up the sea," Captain Wilkinson called it—he was excused many of the more irksome and unpleasant duties normally given to junior cadets.

The most feared person on board the *Marconi* was Miguel the bo's'un, a Portuguese with a scarred flat face and a figure like a Benedictine bottle. Whenever he rolled into view, talk ceased in mid-syllable and the men worked frenziedly. His method of dealing with slackers or troublemakers was simple, yet highly spectacular: he lifted them above his head, carried them forward and threw them against the foc's'le bulkhead.

Until now Bob had managed to avoid contact with this primitive disciplinarian—and, consequently, with the foc's'le's plates. But one morning, emerging from the armory after a particularly successful shark-shooting session, he found the bo's'un blocking his path.

Miguel had his thumbs stuck in his broad leather belt—an accessory which encircled him well below the navel, on the overhang of his great, hairy stomach, and yet by some unknown means defied gravity and kept his trousers up. And he was wearing a thin, cold smile.

"You shoot prettee good, Meester Tuck. What can you do weeth a knife?"

"I've never tried." The bo's'un set off forward, signaling with jerks of his squat head for the boy to follow. They went into the fo'c'sle. On a wooden bulkhead there was a pitted, paintless area, in the approximate center of which a small circle of smudged chalk could barely be discerned in the poor light. Miguel drew a flat, spatulate knife about eight inches long, and with a throw of perhaps twenty feet, from just inside the doorway, buried the blade's point in the dead center of this crude target. Then he retrieved the knife and, with a mocking bow, handed it to the cadet. Some four or five seamen, who had been resting between watches, sat up in their bunks.

Bob let the weapon lie along his right palm. The blade and the slim ebony handle were merged in a perfectly streamlined projectile, beautifully balanced. He had noted the bo's'un's throwing action carefully —the way he'd held the knife, with his thumb pressing the haft gently into his palm; the smooth forward and backward overhand swing, and the powerful flick on release an instant before the elbow straightened and locked. Copying this style, with his first throw he was only two inches outside the circle.

Miguel threw again; another bull's eye. Then Bob split the chalk line. The Portuguese raised his pencil-thin eyebrows and grunted appraisingly. On the third throw there was nothing in it—both just inside the circle.

"You learn queeck, Tommee. Now we throw for a pound, eh?"

Bob hesitated. For him it was a very large sum. But the goading tilt of the bo's'un's head decided him.

"All right. Just the one bet, though."

Some of the seamen got down from their bunks and drifted closer. Miguel was an expert at this game. Until now only he had managed to get inside the circle from this distance. This boy, a proven marksman with a rifle, was the first serious challenger.

Both inside on the first throw. The watchers stirred and murmured. Tuck's second was an unbeatable dead center and even Miguel whistled appreciatively. But on the third Bob was a hair's breadth outside, and the bo's'un's blade thumped home on its mark.

Miguel's smile, as he collected his winnings, was positively affectionate. "You got strong arm, an' good eyes een your head, Tommee. Miguel maybe teach you som' treecks, then you'll be greatest knife thrower een England!"

From that moment on Cadet Tuck was "Portuguese Mick's" proud protégé and only friend. There was no more betting, but almost every day they practiced in the fo'c'sle. The bo's'un taught the boy how to draw the knife from his belt or sleeve and throw it with inch-splitting accuracy in one continuous, lightning movement. And Bob knew he had no more to learn from him when Miguel went ashore at Bahia, Brazil, and returned with a graduation gift—a long, handsome knife in a fine leather sheath.

*   *   *

Tuck's two and a half years at sea—all spent on the *Marconi*—constituted a tougher, more practical—and, as it turned out, infinitely more appropriate—education than his eleven at school. At nineteen he was a lean, bronzed, superbly fit young Englishman. And as a personality he had developed startlingly, mainly because he had learned to dam up his tremendous store of energy until it could be released into useful channels.

In short, he had acquired—or been supplied with—a plan for living, and consequently he surged through each day with confidence, enthusiasm and indomitable gaiety.

He talked a lot, and always at high speed—as if determined to squeeze as many words as possible into each fleeting minute. Though he knew next to nothing about politics, the arts or the sciences, he was perceptive in daily life and therefore knowledgeable about the people he'd met and the places he'd visited. His voyages had provided him with a fund of anecdotes, but he was prone to exaggeration, and held certain violent opinions which he would attempt to ram home with jerky movements of jaw and shoulders. At times, particularly when contradicted or asked to repeat some remark, he had an annoying habit of spacing his words with enforced patience.

In the late summer of 1935, while at home on a few weeks' leave, this volatile and verbose young man happened to spot an advertisement in a daily newspaper and, with characteristic impetuosity, decided to obey its injunction: *"Fly with the R.A.F!"*

It was as simple as that. No hours of solitary thought, no poring over pamphlets, no family conference. Never in his life before had Bob Tuck given a thought to aviation as a career, and no sense of destiny sounded a ghostly fanfare now, as he read and reread the advertisement's rhetorical text.

Though he would have denied it vigorously, though he may not even have been fully conscious of it, his seafaring life was becoming monotonous. At home here he could talk excitedly of savage storms, of romantic islands bathed in the cool, satin light of the tropic moon, and of his wanderings in the tangled streets and alleys of foreign towns. But the more he talked, the more the novelty wore off—the more he thought about the long, uneventful periods in between. Four hours on and eight off, cooped up in a small metal world week after week, doing the same jobs, seeing the same faces, listening to the same voices telling the same old yarns. . . .

The ship seemed to plod her way about the world: his restive spirit clamored for something faster and altogether more adventurous. The air force seemed to provide the perfect answer.

But the air force of those days was a small, exclusive club. Everyone knew everybody else, and most of the senior officers felt it was only a dirty trick played by their landlords, the politicians, that was forcing them to seek new members. (This was the year in which the Air Staff actually opposed the government's plans for R.A.F. expansion, arguing that rapid increase would destroy the quality of the service!)

Though scores of applicants for flying duties were interviewed by selection boards each month, only a very few were accepted for training. Bob's board, held in a lofty Whitehall conference room after a medical and a written examination, was composed of five middle-aged officers. He never learned their names. The chairman, a tall, languorous man, spoke with fastidious intonation and fixed the youth with a bright, unblinking stare. Bob faced him with what he hoped was a composed and enthusiastic expression, and answered his opening queries about schooling, sports, hobbies and the kind of books he'd read.

Then, at the chairman's invitation, each of the other members had a turn. It soon became clear that each was some sort of specialist, confining his courteous interrogation to one particular subject. With the first four there were no tricky or highly technical questions, and Bob was able to give prompt, businesslike replies. It was the fifth and last inquisitor, a bespectacled, fatherly figure who had beamed serenely at him since the start of the proceedings, who with a single polite inquiry pitched him into confusion.

"Tell me, Mr. Tuck," this gentleman coaxed, "what do you know about I.C.E.?"

I.C.E.? Could he mean "ice?" No, that was ridiculous—why should he bother to spell it! The letters must have another significance—but what? "Imperial Chemical . . . ?" Keep talking. Say something—anything. Pretend you understand perfectly—it may come to you suddenly.

"*Well* sir . . ."—a couple of seconds gained by clearing his throat—"I don't think I can claim to know any more about I.C.E. than . . . er . . . the average young man of my age . . . I understand the general principle, of course . . . I've always been interested in it, but . . . but being at sea I haven't had much opportunity to keep up with that sort of thing. . . ."

"Thank you very much," purred his questioner, apparently quite satisfied. With that the chairman brought the interview to an end.

Puzzled and uncertain, Tuck walked up Whitehall to his bus stop. As he watched the bus approach, the truth flashed upon him—I.C.E. meant "internal combustion engines," of course! And mentally reviewing the answer he'd given, he decided he hadn't done too badly. That bit about "being at sea" had been a lucky thought!

All the same doubts lingered until the day, some two weeks later, when he received a letter telling him he had been accepted for flying training for a probationary period with the rank of Acting Pilot Officer, and directing him to report to R.A.F. Station, Uxbridge (Middlesex) by noon on September 16th.

Lamport and Holt accepted his resignation without complaint: his parents complained, but soon accepted it. They realized that their younger son was much too headstrong and eager for adventure to be impressed by their arguments that flying was a highly dangerous, hit and miss profession in which few survived very long. Because he was under twenty-one, they could have stopped him from joining but they knew this would only make him leave home and take up some other, perhaps even more perilous, occupation.

"I'm sorry you're not going on with your cadetship, Robert," said his father. "Seems a pity to waste three years of it. . . ."

"They're not wasted, Dad, I assure you."

Captain Tuck surveyed his son's straight, lithe figure, then looked into the steady eyes above the high, sun-polished cheekbones—and was indeed assured.

"A chap should finish what he takes on. Still, if this is really what you want. . . ." He signed the necessary papers.

Waiting on the platform at Baker Street for the mid-morning Uxbridge train on September 16th, Tuck made the acquaintance of several young men who were to be his close friends and colleagues for many years to come.

"I can see them now," he says, "most of them in blazers and flannel slacks. . . . The first chap I spoke to was Mike Lister Robinson, a helluva good-looking bloke with the sort of fair complexion grocers usually have. I remember he was walking up and down, swinging a tennis racket.

"There was Johnnie Loudon—very tall and so thin and pale you'd have thought he had nine toes in the grave! And Caesar Hull, a little Southern Rhodesian, neat and nimble and tough as nails, with a curious, hoarse voice.

"Then there was Eddie Hollings, always smiling and mad about every kind of sport; his great pal Dougie Douglas, an Australian who'd served a few years in the Navy; Jock Gibson, ginger-haired and very Scottish; "Buti" Sheahan, a strapping South African; Pat Tipping and Laurie Hunt from New Zealand, and Jack Van, a breezy Canadian.

"All told, thirty-three of us got on that train, and all but two got through the course. Today, as far as I know, there are eight still alive."

Uxbridge seemed remote from airplanes and Tuck's conception of air force life. It was big, it was sprawling, it was depressingly bare and old-fashioned. They slept in iron cots in barrack blocks whose bricks had faded. The middle-aged flight sergeants and warrant officers who marched them about the camp and taught them drill were careful to address each "sprog" officer as "Sir" and the whole party as "Gentlemen;" they saluted on the most petty provocation, but nevertheless managed to infer contempt with every glance and utterance. Obviously they were firmly convinced that the R.A.F. would be an infinitely more efficient force if there were no pilots. . . .

Lectures, kitting-up, aptitude tests, parades, P.T., inoculations, documentation—how wistful were the young faces that turned to the sky when, very occasionally, an aircraft droned high overhead. Yet it lasted only two weeks. Then they were posted to Number 3 F.T.S. (Flying Training School) at Grantham.

This turned out to be a big, bleak sweep of Lincolnshire plain dotted with hangars with attractive, comfortable living quarters and a truly magnificent mess building. Above all, it had airplanes—Avro Tutors, long rows of them drawn up on the grass in front of the Flight Headquarters, and several others buzzing fiercely across the patchwork autumn sky.

As soon as they had unpacked their kit they were taken to "A" Flight dispersal. For the first time in his life Tuck stood beside an airplane.

He was shocked. The thing looked so fragile. A light framework of wood and steel covered with thin, tightly stretched fabric and held together by bits of fine wire. The wings were trembling in the gentle wind which breathed across the field.

He touched the side of the fuselage, felt it give a little under the slightest pressure, and immediately thought of the sturdy plates and heavy rivets of the *Marconi*. Somehow he'd always imagined an aircraft to be much stronger in structure, sleeker in shape and altogether more "dependable looking."

The flight lieutenant in charge announced that next morning each member of the course would report for "flight familiarization"—a thirty minute trip, during which the instructor would explain the various controls and how they were used in certain basic maneuvers.

Then he hauled a parachute pack out of the cockpit and proceeded to explain the regulation procedure for bailing out.

A.P.O. Tuck went to bed early that night, and lay awake for a long time, clammily apprehensive.

3

The instructor, Flying Officer A. P. S. Wills, glanced at the cloud-broken sky and said over the speaking tube: "Might be a bit bumpy. Let me know if you feel bad." Then he opened the throttle and the Tutor went bouncing over the tufty grass, vibrating furiously.

In the rear cockpit Tuck sat stiff and unhappy, breathless with the jolting, squashed by the resolute roar of the little Armstrong Siddeley "Cheetah" engine. In mere seconds, they were rocking into the air, the airfield flashing underneath at dizzying speed, sinking away. . . .

There was no canopy—only a tiny windscreen in front of each cockpit. He poked his head over the side to look down at the mess building. A chill, 70-mile-an-hour wind smacked his face and sent a tingle through his body to his fingertips. Almost immediately he knew he would love this.

He relaxed. When they gained altitude it *was* a bit bumpy, but he didn't have a twinge of sickness. Soon they climbed into smoother air.

He felt he was in a dream as they drifted along the morning sunlight and Wills made some gentle turns in the broad blue aisles between the clumps of cloud. He was entranced by the new beauty of the Earth. He looked down on meadows soft and inviting as the evening sea, and was astonished by the geometric cleanness of the farm buildings, the rows of cottages and the great, purposeful latticework of roads and rivers and railway lines that stretched in every direction. It was as if he'd seen the world for the first time.

Wills turned his head and looked at his pupil. The boy grinned back, so he climbed above the biggest clouds and flirted along their sun-touched galleries. Now he made some of his turns fairly steep, and the nearness of the cloud, with the plane's shadow racing over it, provided a strong sensation of speed. From time to time his voice came over the speaking tube— talking about the stick, rudder and throttle and the position of the machine's nose in relation to the horizon during various types of turn, climb and descent. But Tuck was much too excited to concentrate—it all seemed pretty straightforward anyway.

From that first short flight he returned flushed with enthusiasm, soaked in glee. But the following days brought a series of shocks and deepening dismay. For despite his ardor, and the most earnest concentration, he couldn't learn the most elementary exercises— couldn't "get the feel" of the plane, couldn't even hold her steady in straight and level flight. . . .

Wills told him he had two serious, basic faults. Firstly, he was much too heavy-handed and jerky on the controls, overcorrecting drastically every time a wing dipped or the nose rose a fraction above the horizon. He must learn to relax, to move gently and smoothly. Secondly, he seemed quite unable to coordinate the movements of his feet on the rudder with those of his hands on the stick and throttle.

Wills, who had sent many beginners off "solo," and with equal certainty had recommended many others to be dropped from training because they were obviously useless, was deeply puzzled by Tuck. The lad was extremely keen, and certainly not stupid. He knew it all in theory, but for some mysterious reason couldn't get the practical knack.

Wills discussed the problem with his deputy flight commander, Flight Lieutenant Tatnall, and they agreed that the boy was worth a special effort. Their main difficulty would be to keep him confident, to keep him believing that soon it would all come to him—to make him relax, to stop him despairing as the other student pilots went off solo while he still struggled with elementary maneuvers.

And so Wills, by nature a quiet, reserved fellow, went out of his way to chat and joke with his backward pupil in the mess, at flight dispersal, at times even in the air. The friendliness and apparent unconcern of this big, solid man with the crinkly blond hair and freshly scrubbed complexion greatly allayed the boy's fears and increased his resolution.

When Bob made a mistake in the air Wills would say calmly: "Try again, Tucky—that was a lot better." And often, as the student lined up the aircraft to repeat the exercise, the instructor would point downwards and make some cheerful, completely irrelevant remark —"Look at that big herd—Jerseys, I'll bet. One day I'm going to have a farm like that." Or: "See that bloody great pub by the crossroads? The Wheatsheaf. Keep a wonderful bitter there—you ought to nip over and try it one night."

Once, coming in to land, Bob got the nose too high and let the airspeed fall dangerously low. The plane became sluggish, tottering on the brink of a stall. The whistle of wind through the struts and over the wing surfaces died away, but Wills did not take over the controls—he wanted his student to realize what was wrong, and to make the recovery himself. All he said, quite casually, was: "Things seem to be getting a bit quiet, don't they, Tucky?" Bob tumbled to the situation, put the nose down and advanced the throttle just in time to keep her flying. This error was one of the few he didn't make twice.

From time to time he was taken up by Tatnall, and by Flight Lieutenant Lywood, the senior instructor who commanded "A" Flight. Neither of those experienced officers could effect any real improvement, but they too were patient and encouraging. Nothing to do but wait—and hope.

Two of the other chaps on the course were rated unsuitable for flying and "bowler-hatted." Some went off solo after eight hours, others after nine or ten. With over eleven hours of dual time in his logbook, Tuck was still clumsy and quite unsafe. Everyone knew that if he didn't buck up very soon, he would be the third to go back to civilian life.

In the evenings, when they went out in groups to drink beer at the *George* or the *Angel* in Grantham, his friends tactfully strove to avoid talking about their flying. But flying now had become their whole life, and after a few rounds there would come an awkward silence—then, inevitably, they would fall to talking shop.

He managed to hide his growing misery, and none of his colleagues ever so much as hinted that he might be finding things more difficult than they. Yet there was a kind of unspoken sympathy, an atmosphere of goodwill and understanding which told him plainly: "We believe in you, old man—you'll make it all right, you'll see."

On the morning of October 24th, 1935, he awoke depressed and with a slight hangover. At breakfast he attracted some curious glances, but most of the others knew why he wasn't his usual cheerful, talkative self: today he was to be tested by the deputy flight commander, and unless he put up a good show he would be packing his kit before nightfall.

First he had a thirty minute flight with Wills—going over everything he was supposed to have learned. He tried with all his soul to show some improvement, but he gave the instructor one of the roughest rides of his career. Wills' pink face was expressionless as he listed all the mistakes afterwards in the flight hut. "You'll have to do a lot better on the test," he warned.

As he got into the deputy flight commander's airplane, Tuck gave a mental shrug and told himself: "The way I feel now, I haven't a hope. Ah well, the sooner we get it over . . ." He was suddenly utterly weary of the whole business; his mind turned back to his seafaring days, and he thought of his tiny cabin on the *Marconi*—a snug, safe haven he'd be heartily pleased to see again.

Tatnall, a squat and formidable figure in a well-worn flying suit, climbed in without a word and then ordered him to taxi out and take off. Hopelessly, listlessly, he moved his hands and feet on the controls. Tutor K3258 rolled forward, along the edge of the field to the take-off position.

He felt only half awake as he made his final cockpit check. His attention was distracted by a huge thrush yanking a worm out of the ground a few feet from the port wingtip. As he revved up the engine to check his magnetos, the bird took flight and seemed to leap into the air. He watched it climb away in a wide, smooth turn, gaining a couple of hundred feet in mere seconds.

"Yes, it's all right for you," he thought.

And then the speaking tube rumbled.

"You're cleared for take-off," Tatnall was saying. Tuck acknowledged this, then advanced the throttle. His muscles didn't seem to be functioning properly any more—they felt lax and flabby. His hands were no longer able to exert firm grips on stick and throttle, his legs and feet seemed numb and remote—and somehow he didn't care.

Yet, surprisingly, the take-off was almost perfect.

He climbed straight and steady, turned smoothly— and then all in an instant realized, to his complete amazement, that he was flying very much better than ever before! And without any real effort. . . .

In this moment came the great revelation: *flying didn't need any great effort!* You didn't have to think out each physical movement—you didn't even need to make each action deliberate. The whole idea was to make certain hand and foot movements *simultaneously,* in coordination—not separately, as he realized he had been doing up till now.

Provided you knew the principle of the thing, the actual movements should be entirely subconscious. It was remarkably like swimming: you simply made up your mind to turn, or dive, or roll on your back— and automatically your hands and feet applied the appropriate pressures.

A tremendous thrill of victory, and a glorious sensation of power struck a new, hot beat into his pulse. In the next fifteen minutes Tatnall didn't speak except to give directions for a few practice stalls, gliding turns, circuits and landings. Then suddenly they were parked outside flight dispersal, and the instructor was clambering out.

"Off you go!" he said, waved casually, then turned and walked away.

As Tatnall neared the dispersal hut, K3258 was already bumping eagerly across to the take-off point. Wills came up and handed him a mug of tea.

Wills said nothing, but the deputy flight commander saw the anxious look in his eyes and grinned: "He'll be all right—he's suddenly twigged the whole thing."

At 1435 hours, with a total of thirteen hours dual instruction, A.P.O. Tuck flew alone for the first time. He yawned a little on the take-off, but his circuit was neat and correct, and his landing was almost perfectly judged.

At dispersal Tatnall, Wills and Lywood turned as a voice behind them said softly: "Now *there's* a promising lad." It was the Chief Flying Instructor, Squadron Leader W, A. B. "Jimmy" Savile. They'd been so intent on watching K3258 come in that they hadn't noticed his car drive up.

The C.F.I. had a gleam in his eyes as he returned their salute. He kept track of the progress of every student on the station, and he knew very well that it was Tuck, the boy they'd all been so worried about, who had just made that "copybook" touchdown.

He knew it, and he was letting them know he was just as glad as they were that the crisis was over.

* * *

Thereafter, at the end of each stage of training Tuck was graded "above average." Several times he won the rare tribute "exceptional," but the wise and experienced Savile pinpointed his one grave fault when he wrote in his logbook "apt to be overconfident."

That was putting it mildly. Bob reveled in violent airobatics, low flying exercises, simulated forced landings in tiny fields, rough weather flying and "recoveries from abnormal positions." He was a singularly precise judge of height and distance; in low flying and approaches to landing he often cut things so fine that his instructor would brace himself for disaster. His ardent spirit had a zest for danger, and on solos, once well out of range of the airfield, he delighted in illegal hedgehopping sprees and unofficial dogfights with any other students who had nerve enough to take him on. It was all a great adventure—all marvelous fun.

In this era, flying was still a wondrous, hit and miss business with much of the old-time pioneering spirit. The aircraft were light and frail, but they were uncomplicated.

"You felt sure," he recalls, "that if anything went wrong you could pop down into a handy cow pasture and very soon put things to rights with just a little ingenuity, a pair of pliers and your old Boy Scout knife." It was awfully easy to get cocky.

His closest friends were the tough little Rhodesian Caesar Hull, handsome Mike Lister Robinson and Johnnie Loudon—who, though he looked shockingly pale and skinny, was actually one of the most energetic and wildest characters on the course. They lived a carefree and luxurious life—always on whacking overdrafts.

In prewar R.A.F. officers' messes the food was superb. Dinner was never less than four courses—usually five. On party nights, and whenever there was a mess ball, the choicest champagne flowed and white-coated stewards bustled about all through the evening with trays of cold chicken, tongue, smoked salmon and a selection of other delicacies. Everyone would have put on weight had not sports been compulsory.

Tuck joined the fencing club. He was already accomplished with foil and epee, but now he took up saber-fighting and found he liked it best of all. Within a few weeks he was selected to represent No. 3 F.T.S. in the annual Training Command Sports at Cranwell, where he defeated a famous R.A.F. fencing "blue," Flight Lieutenant (now Air Commodore) G. N. E. Tindal-Carill-Worsley.

He also played squash, swam for the Training School in the 100 yards free style event, was a member of the water polo team and—only to be expected—the best shot of the rifle club.

Almost every week-end a merry, lusty band of

youths in blazers and flannels set out from Grantham to take part in some sporting contest against another R.A.F. station or Command. And Bob Tuck was nearly always one of them, for when he wasn't competing he was an enthusiastic, extremely noisy supporter and an eager participant in the beer sessions which invariably followed the official programs.

For instance, he never missed a boxing event, because Caesar Hull was Grantham's best welterweight. "Little Caesar," they called him, and like the emperor of ancient Rome his conquests were legion. Yet out of the ring, you couldn't have found a milder man. You could pull his leg all night and he would only grin. You could be downright offensive, and he would rather walk out of the room than get involved in an argument.

About this time Caesar introduced Tuck to a South African who had just arrived at Grantham in the latest intake from Uxbridge. He was a husky youngster with a broad, open face and lank blond hair. Bob took an immediate liking to him, but they were on different courses so they met only rarely. Nevertheless, this was the beginning of a friendship which was to endure for many years. The blond youngster's name was Adolph Gysbert Malan, but because, like Tuck, he had spent a year or two in the merchant service, everyone called him "Sailor."

They graduated from flying Tutors to bigger, more powerful biplanes—the Hawker Hart and the Audax. Now Tuck's illegal pastimes included flying under bridges, "beating up" railway trains, performing airobatics at a few hundred feet, and power diving vertically from great altitude until the wings began to judder and flap and the needle of the airspeed indicator was trembling well past the "maximum authorized" mark.

Flying alone, often he used to weave dramatic fantasies. A clump of cumulus cloud became a tight-packed formation of fighters with black crosses on their wings—Richthofen's dreaded "air circus." He would dive to battle in company with the ghosts of Mannock, Ball, McCudden and all the other British aces who had fought the world's first air duels twenty years before. And in such moments he would sometimes curse the fact that he had been born too late to share the glory of "the real thing."

One day when he came back from what was supposed to have been a cross-country solo in a Hart, Wills picked a piece of greenery out of the undercarriage.

"How did this happen?"

"Got lost, sir," Tuck said blithely. "It was very hazy, so I had to go down pretty low and read the name of a railway station. There were a lot of big trees, and I suppose I must have nicked one . . ."

Wills fingered the greenery and looked at him for a long time before he spoke again.

"You've got to ease up, m'lad. You're too damned sure of yourself. Yes, I know you're hot stuff—but I want you to remember this: nobody ever gets to know everything about flying *Nobody*—understand?"

"Yes, sir."

Wills stared at him for a moment, then tossed the piece of greenery on the ground and walked away. At the time Tuck wasn't impressed, but the day was to come when he realized that Wills' words constituted just about the best advice any experienced pilot ever gave to a beginner. He was to pass on this advice to scores of youngsters during the Battle of Britain.

Wills almost hoped that Tuck would have a crash— a minor one, of course, just enough to give him a sharp scare and show him he wasn't indestructible. It would be better for him to have a bit of a scare in a Tutor or Hart now, so that he would have some of that cockiness shaken out of him before he went on to the faster Fury fighter. But Bob's aircraft handling was too perfect: with two-thirds of his training completed, he was one of the very few students who hadn't by his own error inflicted some slight damage on a machine —not even a scraped wingtip, a strained undercarriage, or a bent tailskid!

Nevertheless, his downfall came. Not in the air—it all happened miles from the flying school, on the ground. On the main Grantham-Lincoln highway, to be exact.

He and Johnnie Loudon were hellbent for Lincoln in Loudon's dilapidated Morris 8 saloon, having spent three hours or so shaming the regular market day drinkers in the *George* at Grantham. Johnnie tried to round a corner at nearly sixty, went off the road on the left side and up a steep, grassy bank. The car tilted to the right, rolled clean over twice and came to rest upside down on the opposite side of the roadway, a matter of feet from the front door of the Braistbridge Heath Police Station.

The station was in fact an ordinary cottage—one of those poky country establishments manned by a single constable. To this day Tuck is convinced that, plastered as they were, this officer might have let them off very lightly—if only they hadn't crawled out of the wreckage roaring with laughter, and slapped his back so resoundingly . . . if only they hadn't clumped about in the kitchen making such a din that they wakened his wife, asleep upstairs, poor woman, after having gallstones removed . . . if Johnnie hadn't decided to try on their host's helmet and render a chorus of "We Run Them In. . . ." Above all, if one of them—he refuses to say which—hadn't wandered out into the back garden and watered the constable's prize cabbages so thoroughly. . . .

The upshot was that while they were washing off the dirt and oil, and doctoring their few minor cuts and abrasions, the constable was on the telephone to his inspector at area headquarters. In a very few minutes the place was blue with policemen, and very soon after that A.P.O.'s Tuck and Loudon were cautioned and charged, most uncommonly, with "being drunk on the King's highway."

A permanent staff R.A.F. officer was fetched to escort them back to Grantham, where they were immediately placed on "List B." This was much the same as being put under open arrest: they were confined to the station and obliged to report to the duty officer at regular intervals throughout the day and night, wearing full kit.

For what seemed an eternity—it was in fact only a matter of a fortnight or so—they remained on "List B," and at the same time continued with their normal training duties. Then one morning they found themselves standing in the dock at Kesteven County Court.

They pleaded guilty. The clerk of the court inquired if there was a senior officer present to speak on behalf of the accused. A massive figure arose at the back of the room and boomed: "Yes, I'd like to say a few words."

It was "Big Bill" Staton, a wing commander from the administrative staff at Grantham—one of the most famous pilots in the air force. Staton had started on Royal Flying Corps fighters in the Great War, and shot down twenty-seven German aircraft. Since 1919 he'd flown everything from bombers to seaplanes and the new flying boats, broken several air records, fought rebel hillmen on India's Northwest Frontier, tested dozens of experimental aircraft and conducted an exhaustive aerial survey of the Far East.

Tuck and Loudon watched with worshipping eyes as Staton—well over six feet tall, and a good three across the shoulders—lumbered forward to the well of the court. His "few words" turned out to be a detailed

account of each young officer's progress since joining
the service. He even included the gradings they had
received from Savile—"above the average, excellent."

He emphasized that neither man during his air force
career had been the subject of any formal disciplinary
action. He reminded the court that the R.A.F. selected
its candidates for flying training from many thousands
of applicants in all parts of the Empire.

"Those who are accepted," he declared, "are indeed
the flower of British youth."

Finally he suggested that the accused—"both mere
lads, still in their teens"—were not used to alcoholic
beverages except in very small quantities. *And this he
said with a straight face, looking the magistrate in
the eye. . . !*

The flower of British youth were fined ten shillings
apiece. Outside the court they waited for their deliver-
er, and Tuck stepped forward as he appeared.

"We'd like to thank you, sir. It was terribly decent
of you to come all this way and . . ."

That was as far as he got. Staton's great jaw jutted
out till it seemed to cast a cold shadow over both of
them.

"Don't give me that b-loody bullshine!" he bellowed. "If I'd had anything to do with it the pair of
you'd be on bread and water for a b-loody month and
b-loody well grounded for six!

"You think you're too b-loody good. It's types like
you that waste the air force's time and money. You
take everything it has to offer, and give damn-all in
return. I hope to hell I never set eyes on either of your
ugly b-loody mugs again." He almost bowled them
over as he strode off.

All the same, they were relieved that it was over.
Back at Grantham, they were taken off "List B," and
life returned to normal.

One afternoon Tuck was lolling on the grass outside
the dispersal hut idly watching a Fury come in to land
when Caesar joined him.

"Know who that is?" asked the Rhodesian, dropping down beside him. "Your old chum 'Portia' Staton.

Somebody just told me this is only about his second or third trip in a Fury."

Bob propped himself up on his elbows and watched more attentively as the biplane glided smoothly over the boundary fence.

The aircraft flattened out perfectly and skimmed along, gradually losing speed, a few inches above the grass. They could see Staton's great bulk rearing out of the cockpit: he was heaving himself up and sticking his head out over one side so that he might still have a clear view when, in the next few seconds, he eased the machine's nose up steeply for the touchdown. As he drew nearer they saw that his face was bright red. It was quite a comic spectacle—the machine seemed several sizes too small for him, and his whole posture was eloquent of acute discomfort and mounting irritability.

"I say, he really is in a fury!" Caesar said, and they both started to chuckle.

The plane's tail dropped lower, and she sank gently to the grass in a perfect three-point position. But as soon as the wheels touched, the tail whipped up violently. The propeller blades churned up showers of dirt and clods of earth in the instant before they snapped off. Over she went in a fierce, full somersault, and came to rest on her back with the whole structure of wings, struts and wires spread around her in crumpled confusion.

Bob and Caesar sprinted out and reached the wreck in a matter of twenty seconds or so. The wing commander was trapped in the cockpit, upside down. Because of his great height his head was pressed against the ground, and the weight of the aircraft was forcing his chin forward on to his chest, so that his mouth and nose were buried in the thick fur collar of his flying suit. To add to his plight, he had bashed his face against the instrument panel and his nostrils, mouth and throat were choked with blood. If they didn't get him out very quickly indeed, he would certainly suffocate.

There was, of course, imminent danger of fire and explosion. Gas was pouring out of the smashed engine and hissing angrily on the hot exhaust pipes. Caesar threw himself flat, wriggled underneath and clawed the fur collar away from Staton's face. Then he reached up and tore at the fastenings of the safety straps. Meanwhile Tuck got his shoulders under the tailplane and, heaving with all his might, managed to raise the tail about a foot. This brought the injured man's head just

clear of the ground. Caesar got the straps loose and was able to drag him halfway out of the cockpit. Then Tuck let the tail down again, and ran round to help. With the aid of others who now arrived on the scene, they hauled him well away from the wreck.

Staton's eyes were bulging, he was blowing large red bubbles and making loud gargling noises. They thought they might have to give him artificial respiration, but after a moment he sat up, spat out a lot of blood and began to breathe more easily. He was still very dazed, but able to get to his feet. Leaning on somebody's arm, he walked to the ambulance and was whisked to sick quarters.

When he'd been patched up, "Big Bill" asked to see the two officers who had been first to reach the wreck. But when they told him A.P.O. Tuck was one of them he said: "My God! Never mind—just tell them I'm very grateful for their help. That'll have to do." After writing off an aircraft worth thousands of pounds, he simply could not bring himself to say "thank you" in person to the youngster he had upbraided so recently for "wasting the air force's time and money."

(The cause of Staton's accident was no mystery. Everyone on the station knew what had happened: as he heaved himself up in his seat to get a clear view for landing, he must have braced his feet against the Fury's rudder pedals, forgetting that by depressing the pedals he was operating the undercarriage brakes. In all fairness it must be said that he was used to bigger aircraft, with different braking systems, or no brakes at all. At any rate, it seemed clear that when the fighter touched the grass the wheels were locked rigid—and base over apex she went.)

A few weeks after this incident, Tuck and his fellow students qualified as pilots and received their wings. Their probationary service was almost over—in another month or so they would be pasted to squadrons as fully fledged officers. In the meantime they moved to North Coates Fiddes, in Lincolnshire, and went on to the final and most exciting stage of their training—gunnery.

From the very start Tuck, the experienced marksman, excelled in both air to air and air to ground firing. The Fury was armed with two Vickers .303 machine guns. These were fixed to the fuselage forward of the cockpit, and fitted with the Constantinesco synchronization device which made it possible for their streams of bullets to pass through the whirring propeller's arc without damaging the blades. (This device went wrong now and then, causing considerable inconvenience to pilots and heavy expenditure on spare propellers.)

The commanding officer of the unit, Squadron Leader B. H. C. Russell, graded Tuck "above the average" and wrote in his confidential report: "This officer is a born fighter pilot."

The course at North Coates Fiddes ended with a popular and highly spectacular event known as the air gunnery competition. Though in fact it wasn't officially recognized as a contest, but as a passing-out exercise, it was the basis of a huge sweepstake organized by the ground crews. There were few instructors in the whole of the Command who did not have bets on the result, and most of the senior officers—it was rumored even Group Captain Nutting—usually put on a pound or two as well.

This time Tuck was the clear favorite, and his nearest rival was Loudon—who, oddly enough, had done very little shooting until he came into the service. But, just ten days before the competition each received an official letter from Air Ministry:

Sir,
   I am commanded by the Air Council to inform you that they have seen fit to terminate your temporary commission. . . .

They were being "bowler-hatted."

It was a profound shock—beyond all comprehension. They had been punished once for their drinking spree—by a civil court, and by the weeks of anxiety and "List B" indignity before the trial. And they had been permitted to go on and win their pilots' wings. Now, when the whole affair was almost forgotten, the Air Council had decided to impose a second, and incredibly drastic, penalty. If their commissions had been confirmed they would have had the right of appeal—right up to the Sovereign. But as probationers there was nothing they could do.

To Tuck it was like waking up and finding himself orphaned. It was getting on for four years now since he had lived with his parents; the air force had become his home, his family. (Even Aksinia, for whom he had always felt a deep affection, had left for her home in Russia, and her letters had petered out some months ago.) As he packed his kit, for the first time since early schooldays he was close to tears.

A mournful throng gathered at the main gates to bid them farewell—Caesar Hull, Gibson, Robinson, Hollings, and the rest. They shook hands, scuffed their feet and mumbled self-consciously. Hollings asked: "Where are you going? What'll you do now?"

"We're going to London," Johnnie said, "and get screechers again—only more so. And when we've had enough we're going to blow up the bloody Air Ministry."

In London they did get plastered, but the more they drank the more their feelings of injustice and desire for vengeance gave way to self-pity and sheer, inert misery. The problem now was how to cleanse their own consciences, to regain their self-respect. It could not be done unless they could find a way to continue flying.

Rumpled heads drooping over their beers, far into the Mayfair night they discussed the possibilities of getting jobs with some air line . . . of joining the Brazilian or the Chinese Air Force . . . of stealing an aircraft from Hendon or Croydon and flying it to Spain or Abyssinia. . . . In the end they decided to form their *own* air force—they would enlist in the Foreign Legion with *carte blanche* from the French government to inaugurate a "Desert Flying Unit."

In the sharp, vibrating light of morning, with only shillings left, all their defiant resolution evaporated. Johnnie summed up courage and set off for home, but Tuck wasn't ready to face his father yet—he must have time to think out the best way of breaking the shameful news. He collected his gear and took a train to Kingswood, in Surrey, where his favorite aunt and uncle lived.

His mother's youngest sister, Rosalind, and her easy-going Scots husband Robin Hall-White listened to his story sympathetically and agreed to keep it all quiet until he was ready to face Captain Tuck. They gave him the guest room and that evening, in an effort to cheer him up, took him out for a few beers at their local. When they returned home a little after nine, a small yellow envelope lay on the doormat.

The telegram was addressed to R. R. S. Tuck. He tore it open with fumbling fingers and read: "Report for duty soonest possible everything straightened out regards (signed) W. A. B. Savile. Squadron Leader."

He threw his gear into his suitcases. Uncle Robin drove him at top speed to Liverpool Street and he caught the first train north. As he walked into the mess building at Grantham a wild figure charged to meet him, arms outstretched, making shrill squealing noises —Johnnie Loudon, also reinstated by telegram. They crashed together, hugged like wrestlers and executed a merry jig.

"Lucky bastards," Caesar Hull said when they rejoined the others at North Coates Fiddes. "You're only back for one reason—because Groupie Nutting and a lot of other brasshats have backed one or the other of you for the shooting competition!"

The truth—as they discovered in due course—was that Lywood, Savile and several other instructors had got together and appealed to the Group Captain, and through him to Air Ministry, not to dismiss "two of the most promising young officers in the service." After examining their flying records and grading, Air Ministry had relented.

Tuck came first in the shooting competition, Loudon second. In the circumstances, it seemed there was nothing else they could do. . . .

But when the squadron postings came through, Tuck was sent to 65 at Hornchurch (4), and Loudon to No. 23 at Northolt (3). Apparently Britain's defense policy demanded that in future they should be kept well apart.

4

It should all have finished on January 18th, 1938. Finished in a mound of mangled metal in a meadow near Uckfield, Sussex—ingloriously, stupidly, wastefully. Luck—the fantastic luck that began that day, and never deserted him through his entire flying career—was what saved him.

They were flying Gloster Gladiators now, tough little biplanes with powerful radial engines and enclosed cockpits, equipped with radio and armed with four .303 machine-guns. (Gladiators remained in service long after the outbreak of hostilities, and flew with surprising success against enemy monoplanes of infinitely superior speed and armament in Norway and Malta—where for a period the battle was fought by just three of them, the famous "Faith, Hope and Charity."

In a way, Tuck *did* die that day—that is the dogmatic, cocksure, incorrigibly reckless Tuck that 65 Squadron's senior officers had long marked down for a sticky end. The Tuck that survived came back to Hornchurch with a livid scar down his face, and a new personality—a more mature and reliable officer, altogether more likable.

So here let us pause, for a brief obituary to the old Tuck, the carefree youngster who clambered into his plane on that January afternoon, so sure that the sky would never do him harm.

He had only one great fault in his flying—he was *too* brilliant. Since his first day at Hornchurch, eighteen months before, he'd had only one object: to be the top man in every branch of flying. And, grimly determined under his veneer of casualness and gaiety, he had succeeded in leading all the junior officers in everything from airobatics to radio navigation, from formation flying and interception to ground attacks and reconnaissance.

He was fastidious in all his work. If he thought he was getting slightly ragged or rusty at some particular maneuver or exercise he would give up his day off to practice and polish. He got his fitters, riggers and electricians to show him how to take his whole airplane to pieces and reassemble it. He devoured technical books and magazines and the biographies of the world's greatest fliers, made long journeys during leaves to visit aircraft factories and talk with the design and test staffs.

Often when it poured and flying was "scrubbed," he would don an old sou'wester and go alone to the station's skeet range to spend two hours or more shooting at fast-moving clay pigeons launched from a springloaded catapult by a soaked and sullen "erk."

Once he kept the squadron's engineering officer up till dawn arguing the technical points of a new type of airspeed indicator. He claimed to have spotted a flaw—before it had even been tested in squadron service. The E.O. called him "a pig-headed young fool" —but had to apologize a couple of weeks later when the new instrument went back to the makers for drastic modification.

There can be little doubt that Tuck's fervor sprang from wounded vanity. He could never forget his first days at Grantham, when he was "dunce" of the course and thought he'd never get off solo. Now he had decided—subconsciously, possibly—that if he could remain consistently ahead of his colleagues, in time this memory would cease to haunt him.

The only young officer who came near to his standard was a member of 74 Squadron, which shared the aerodrome with 65—Caesar Hull's chum from Grantham, Pilot Officer "Sailor" Malan. Rivalry didn't prevent Bob and Sailor from becoming very close friends; indeed, it was Sailor who now revived the old *Marconi* nickname, "Tommy Tuck."

In many ways they were precise opposites. Malan was fair, thickset, relaxed, not too talkative; Tuck was dark, thin, highly strung and gregarious. The South African couldn't hum three notes without changing key, while Bob was developing a wide knowledge of classical music and turning over a fair whack of his pay each month to share in the hire-purchase of a radio phonograph.

Sailor, in fact, seemed to sail through life. He didn't get even mildly excited, never took unwarranted risks, came out tops every time without really trying and then got embarrassed if anyone said "good show." Bob was intense, ever-striving, a hard worker and a hard player—and the more successful he was, the cockier he became.

But they had in common a love of flying, all forms of shooting, bitter beer and fast sports cars.

In this period he made a number of other friends, on both 65 and 74—pilots who were to fly with him in the war years. Among those were "Chad" Giddings, Johnnie Welford, Jack Kennedy, George Proudman, Norman Jones, Gordon Olive, Peter Hillwood (who was later captain of the record-breaking Canberra) and "Nicky" Nicholas. The few who are alive today—of the above, only Jones, Hillwood, Olive and Nicholas—still keep in touch by letter, or by attending service reunions once or twice a year.

Tuck refused to admit that any of his flying was in the least dangerous. He had such confidence in his own skill that he once told his flight commander, Flying Officer Leslie Bicknell: "Some of the things I do may *look* a bit shaky, but that's only because I'm able to judge everything so finely."

Bicknell must have been convinced, for he chose him to fly as one of his wingmen in a public display of airobatics on Empire Air Day. Three of them in "vic" formation performed loops and rolls, and flew upside down a few feet over the upturned faces of a huge crowd, their wingtips linked by twenty-foot lengths of thick cable. The third pilot was a cheerful, good-looking young sergeant named Ronnie Morfill.

The thought that he might actually crash was never offered a moment's lodging in Tuck's mind. He was utterly confident it couldn't happen to *him*—and anyway, pilots in those days often wrecked their machines and escaped unhurt. This latter theory gained much credence from the adventures of one Samuel "Fishy" Saunders.

A gangly, awkward lad on the ground, Flying Officer Saunders was generally reckoned a pretty fair pilot. But one night he completely misjudged a landing approach, undershot the flare-path—to be blunt about it, undershot the airdrome!—ploughed across a bumpy field and shattered his Gladiator against the stout stone wall of a farm outbuilding. Very chary of fire, Fishy scrambled quickly out of the débris, dazed but without a scratch. Then, as he started hurrying away, he trod on a hunk of buckled metal and sprained an ankle. . . .

This experience struck the other 65 boys as extremely funny. But it also served to confirm Tuck's dangerous theory that if you *did* smash up a kite—through somebody else's error, of course—the chances were that you'd walk away from the wreckage, hooting with laughter.

Even at this early stage, Tuck had formed very definite opinions about fighter tactics. He was, for instance, convinced that the tight, "guardsman" formations they were made to fly were a waste of time. The pilots were too busy watching each other's wingtips to keep a proper lookout for enemy aircraft, and they consumed vast amounts of fuel by juggling their throttles to keep station. Besides, a group of fighters packed close together provided anti-aircraft gunners with one large target instead of several small ones, and were much more "solid" in appearance and easier to spot from great distance.

He was alarmed by the number of foreign military missions courteously received by the R.A.F. and shown "the works." A party of Germans, led by Luftwaffe generals Erhard Milch and Ernst Udet, Great War aces—both of whom later became field marshals —had visited Hornchurch soon after 65 was equipped with Gladiators. The pilots had been warned by the station commander, Group Captain "Bunty" Frew: "You can answer any questions they ask you, except about defensive tactics, operations control and the new reflector gunsight. If they get on to the gunsight, tell them it's so new that you haven't learned how to use it yet!"

The reflector sight constituted an important advance in military aviation. The pilot looked through a small piece of plate glass, on to which a system of lenses threw a faint red circle with a dot in the center. Across this circle two horizontal lines were superimposed, on a level with the center dot. These lines could be moved inwards or outwards by turning a milled ring at the base of the sight. The milled ring was graduated to represent the wingspans of various types of aircraft. Once the wingspan had been set, the pilot knew he would be within range to open fire when his target filled the center gap between the two horizontal lines. A small brilliance control was fitted, allowing the pilot to adjust the intensity of the reflection for various combat conditions.

The aircraft were lined up on the field with the pilots standing beside them, and the Germans came round to inspect accompanied by an interpreter and an Air Vice Marshal. When they reached Tuck's aircraft, sure enough Milch got up on to the wing, stuck his head in the cockpit and inquired in passable English how the gunsight was operated. Bob stammered out the prescribed answer and the German, smiling faintly, started to turn away. But at once the Air Vice Marshal said: "Allow me, General!"—and jumped up beside him.

For the next few minutes Tuck stood by, appalled, while the brasshat proudly and lucidly explained the principles and advantages of the new instrument, and even demonstrated the various settings. He was sorely tempted to suggest: "Sir—why don't we give General Milch one to take home, as a souvenir . . . ?"

Apart from his knowledge of international affairs in relation to military flying, Tuck interested himself scarcely at all in the events of the outside world.

His reading was still almost completely limited to text books, technical magazines, the lives of famous aviators and the more popular daily newspapers. But from time to time, in a rare mood of calmness and insularity, he would shut himself away for several evenings and wander, half lost, in the mighty works of Tolstoy, Gorki, Chekov, and Mikhail Sholokov. *War and Peace* he waded through three times. Pre-revolution Russia always fascinated him—this was his only true escape from workaday reality, a passion kindled by Aksinia's romantic tales years before.

The movies, on the whole, bored him. He was seldom convinced by any actor's performance—he could never quite forget the mechanics of projector and screen, and this made a silly charade of the whole business. In particular, he had a fierce scorn for pictures about flying, because in several which he'd gone to see in a spirit of genuine enthusiasm and even excitement, he'd spotted glaring technical errors, fallacious escapes in the plot, and flight and combat scenes which to his trained eye were obviously faked.

And yet, one sort of film held strong appeal for him. He would drive many miles to see a good Western. His favorite stars were William Boyd, Buck Jones and the up-and-coming John Wayne. Though he couldn't lose himself in the picture—couldn't make himself believe for a moment it was anything but a piece of elaborate pretense by overpaid, soft-living "theatricals" —he thoroughly enjoyed the grandeur of prairies, mountains and deserts, admired the undoubted skill of the horsemen, and shook with laughter over the miraculous shooting of the hero. His whooping mirth caused considerable annoyance to people sitting near him—as did his habit of counting aloud the number of shots fired by a selected individual in the course of a gun battle. He would go about the mess for days afterwards telling everyone how one of the actors had fought his way out of a tight spot by firing eleven rounds without pausing to reload his six-shooter. . . .

Several of the younger pilots were keen on jazz and dance music. To Tuck it was a meaningless din. Nor could he derive any enjoyment from light music or opera; for him the only true music was that of the Masters—Bach, Beethoven, Brahms, Mozart and Wagner—the sterner the better. It is difficult to see how he had developed such taste; although his mother was an accomplished pianist, he'd had no musical education.

It was partly his views on music, and partly his lack of interest in women, that kept him rooted to the bar at mess dances. When absolutely necessary, he could shuffle rigidly through a foxtrot or waltz—*preferably* a waltz—with the wife or daughter of a senior officer. In close proximity to young femininity he invariably became fidgety and confused. He wasn't afraid, he just didn't know how to begin making conversation with a girl—she was a creature from another world, with whom he had absolutely nothing in common. His biological instincts were quite normal, but at this period of his life they were sublimated: all his extraordinary energy, his spirit of adventure and his masculine pride were directed to a single end—to keeping his place as the squadron's best pilot.

*   *   *

So much, then, for the carefree, headstrong, barnstorming Tuck, to whom we now must say farewell. For no amount of skill and judgment, nor soaring confidence, could have saved him on January 18th—though let's get it perfectly straight, the accident was not his fault.

They were practicing formation at 3,000 feet over Sussex—Tuck, Flying Officer Adrian Hope-Boyd, and a new sergeant pilot named Geoffrey Gaskell.

Though he considered tight "text-book" formations thoroughly bad policy from a military point of view, Tuck enjoyed flying them. With chaps like Bicknell and Hope-Boyd, he found it exciting to see how close they could get—to edge in until their wingtips literally overlapped, and to hold their positions, steady like shadows of the leader's machine, while they performed steep turns, dives and complicated battle moves.

But today was different. There was a high wind and from time to time they flew through turbulent air which rocked their wings. Besides, Sergeant Gaskell had very little experience—in fact, Bicknell had ordered the flight for Gaskell's benefit, to give the N.C.O. a chance to practice with two of the squadron's best men. So they kept it fairly loose.

Somewhere near Uckfield they moved into line-astern —Hope-Boyd leading, Gaskell next, Tuck last. Each of the last two planes was slightly higher than, and a shade to the right of, the one in front—just enough to stay out of the slipstream: at this close range, the invisible wake left by a Gladiator's propeller could buffet as violently as rollers churned up by a steamship's screws.

Suddenly they hit a patch of rough air. Gaskell's machine bucked and he must have overcorrected, for he seemed to veer to the left and then drop back down too far, directly behind Hope-Boyd. The leader's slipstream caught him and threw him over into a steep bank.

All this occurred in the space of about three seconds and Tuck, behind and slightly above Gaskell, was poised to yank the stick over and get out of the way if the number two plane didn't settle back into position very quickly. He was actually easing off his throttle and beginning to fall back slowly when Gaskell, who could have broken away to the left into empty sky with complete safety, suddenly, unaccountably, hauled his plane up on its tail and broke *to the right*.

Bob saw him rear up, dead in front. There was no hope of pulling sideways or upwards to avoid collision. For one horrifying instant, over the top of his own engine cowling he looked down into Gaskell's cockpit . . . saw the sergeant's gloved hands frantically working stick and throttle, saw the top of his leather helmet pressed back rigidly against the headrest, the folded map protruding from the big patch pocket on the left leg of his flying suit. . . .

The two machines were almost at right angles. Instinctively Tuck slammed the stick forward, forlornly hoping he might still be able to dive under Gaskell's tail. But his nose had hardly begun to go down when they struck.

The noise was like a splitting oak. Tuck's propeller sheared through the fuselage of the other plane close behind the canopy, and almost certainly sliced into Gaskell's head and back.

Beyond Bob's windscreen there was nothing now but writhing sheets of fabric and buckling metal airframe. The wings crumpled and wrapped themselves round his hood, enveloping him with the finality of a shroud.

He waited for the engine to burst—for hunks of jagged mechanism to hurtle back into the cockpit, smashing him to a pulp. He waited for the first flicker and crackle of the fire which would cremate him as he sat here, sealed in his tiny room, high above the Sussex farmlands.

But the engine did not burst and the flames did not appear. Slowly it dawned on him that the engine must have fallen outwards, for now the only sound that reached him was a hideous grinding and flapping from the tangle of smashed wings over his head. He couldn't see out, because wreckage from Gaskell's aircraft was still draped across his windscreen, but somehow he realized that the machines had broken apart now, and that what was left of his was hurtling earthwards.

Then he began to fight. He got out of his straps and disconnected his radio wires and oxygen tube easily enough, but when he tried to slide back the canopy he found it would not budge. The mess of crumpled wings outside was holding it tight, immovable, like the lid of a coffin.

A singing in his ears and a growing pressure behind the eyes told him that the wreck was spinning faster and faster. He smashed at the canopy with his fists, tore at it with his fingers until the nails were ripped and sticky with blood. But the canopy remained solid, unyielding.

He stood on the bucket seat, bent double, got his shoulders under it and heaved upwards and backwards with every fiber straining. He might as well have tried to lift St. Paul's Cathedral.

To get a better purchase, and in order to push in a directly backward direction, he placed both feet on the instrument panel: he still retains a vivid mental picture of his feet disappearing through the "dashboard" in a tangle of wires, smashed glass and splintered paneling. Then contrifugal force threw him to the bottom of the cockpit, a flailing, kicking heap.

It was now that the terror came. Like a mist, blinding and choking. It bit coldly into his belly, knotted his muscles and set his whole body aquiver. It crushed him, like an assailant kneeling on his chest. Stampeding blood clouded his brain, and suddenly he was yelling and cursing in a strange inhuman voice.

Here Fate relented. As the wreck's spin grew faster and tighter, there was a deafening groaning and flapping, and the mounting stress and wind force tore away the remains of the wings which held him prisoner. Daylight flooded the cabin again, and its glare shocked him out of his flailing panic.

He hauled himself forward, reached once more for the handle of the canopy. Through the windscreen he had a momentary impression of the ground rushing towards him and thought: "It's too late now." His numbed fingers couldn't find the handle. It took him precious seconds to understand why: the canopy was gone too. . . .

As his head and shoulders emerged, a giant, invisible hand seemed to grip him and press him backwards, pinioning him against the rear of the cockpit. He kicked and struggled furiously, and after a few seconds felt himself plucked out of the wreck like the cork from a bottle. In the sudden, almost weird silence he seemed to hear the thought shrieking through his brain: *will my 'chute open in time?*

His hand flew to the ripcord, fumbled, found it, yanked desperately. A field with a thicket of trees at one end rose vertically in front of him like a blind, then slid over his head as he fell into a slow, forward somersault. He waited, lying spreadeagled in the sky, giddy and sick and ridiculous in his helplessness. Hope drained away with every instant, until he was convinced that the parachute had failed.

A plop and a sharp tug, and the world had been put to rights. He was on an even keel again, floating tranquilly downwards through the sunlight, with fully three hundred feet to spare and open grasslands beneath. The surge of relief was like the enveloping caress of a warmed bathtowel. He let himself go limp and sobbed in great lungfuls of air. He wanted to laugh—or perhaps to cry, he couldn't be sure which. It was exactly as if he were drunk.

Then he became aware that his clothing was wringing wet, and thought "I must have sweated buckets." He looked down at the front of his white, flying overalls. It wasn't sweat that made them stick clammily to his chest—it was blood. Lots of it. As his fuddled brain strove to cope with this new situation, he saw crimson spurts shooting down—it was coming from his throat!

Frantically he tore the silk scarf from his neck. As his fingers groped for the wound, the blood flowed so heavily that it streamed down his forearms and cascaded from his elbows. An artery must have been severed—he must pinch it closed with his fingers quickly, or he would surely bleed to death.

The strange thing was, he had no pain in this region, no recollection of receiving a blow.

He couldn't find the wound: under his chin everything was so soaked and slippery that he couldn't tell whether the flesh was whole or torn. As he put his head back he began to cough, and then the blood started to come from his mouth and nose. He wiped some of it away with the back of a hand and found that the right side of his face seemed to be hanging loose and flabby. Before he could explore further, the ground came speeding up to meet him and he had to brace himself for the landing.

The impact was harder than he expected. One ankle twisted awkwardly. As soon as he had spilled the air out of the silk and straightened to a sitting position, his hands flashed to his right cheek. Two of his fingers went right through the side of his face and into his mouth. A back tooth rocked precariously under them. He gripped it gently, and lifted it out through the gaping hole.

Probing cautiously, he found that the cheek had been laid open from the top of the ear almost to the chin. The cut was dead straight. He would never know for certain, but it seemed very probable that as he was struggling out of the cockpit a loose piece of wire strutting—knife-edged, to cut down wind resistance—had lashed his face. It had missed his eye by little more than an inch.

He managed to staunch the flow to some extent with his scarf, but when he tried to stand up the field tilted crazily, and his twisted ankle gave way. After that he just lay back in the grass, cursing softly, until some farm workers came running to his aid.

He left hospital six days later with a slight limp, the long and permanent scar on his face—and changed ideas. He'd lost a back tooth, but found a little wisdom.

A Court of Inquiry absolved him from all blame for

the accident. It was stated in evidence that the front half of Gaskell's machine had plummeted into the grounds of a nursery at Ridgewood, Sussex, wrecking the end of a barn. The sergeant was still in the shattered cockpit, but he'd been dead long before that final dive to earth; medical witnesses said he must have been killed instantly as the two aircraft collided.

Tuck's machine was found in a field over a mile from Ridgewood. It still had the tail and part of the fuselage of Gaskell's plane attached to it—jammed between the legs of the fixed undercarriage.

Now Tuck's style of flying—he was back in the air just nine days after the accident—altered markedly. His nerve remained as steely as ever, his judgment precise and his enthusiasm boundless, but he no longer took deliberate risks just for the thrill of it. He knew that only luck—not skill, not daring—had saved him from a miserable death, and he had learned that in military flying there were unpredictable factors which killed the best and the worst pilots with terrible impartiality.

And now he knew, too, that he was not immune to fear.

It would be untrue to say that he became, even for a period, a "careful" flier. Qualities like prudence, meticulous observance of rules, restraint and resolution go to make fine bomber captains, but the good fighter pilot must have a touch of devilry that derives joy out of throwing away the rule book in order to attempt that which is forbidden or supposed to be impossible. With Bob Tuck, this inner sprite was not strangled, but merely disciplined. So it was that he developed, at this very early stage in his career, something extremely rare, something which usually comes only after many years' experience: a kind of *controlled* recklessness, based on flying knowledge, and indulged in only when there is something worthwhile to be gained.

In an effort to distract attention from his scar, he now grew a mustache. Regulations said mustaches should be full—untrimmed. But Tuck's lean face couldn't support a bushy growth—he looked "like a greyhound with antlers." So he clipped it to a slender black bar. Nobody reprimanded him.

He was involved in only one more accident during his pre-war service—another air collision in April, 1938. Practicing airobatics with Leslie Bicknell 6,000 feet over Great Warley, Essex, he was edging in close on the right of his leader when suddenly another "vic" of Gladiators came shooting up out of a bank of cloud very close underneath, directly ahead—a chance in a million. Bicknell was forced to bank sharply, open his throttle wide and go into a steep climbing turn to the left.

Tuck tried to follow, but because he was a second or so later in opening up to full power, he couldn't get on top of the turn. His lower port wing struck the rear of the leader's machine, and to his horror the rudder, fin and tail plane broke off.

Immediately Bicknell rolled on his back and went into a flat spin. Tuck dived after him, and only then discovered that his own machine was badly damaged: almost half of his port mainplane was smashed and hanging downwards at right angles, and the rest of it was buckled and liable to collapse at any second.

But he went down after Bicknell all the same, praying that he would see the white bloom of a parachute —that Bicknell was not trapped in the cockpit as he had been three months before. . . .

The spinning wreck sank into the mattress of cloud and he lost it, so he dived on through and waited, sick and breathless, underneath. After what seemed a very long time the tailless fighter came spiraling down out of the murky overcast. It was two or three hundred yards away, and in the poor visibility he couldn't be sure whether its canopy was open or closed. He flew towards it, but before he could get near it struck the ground in the middle of what looked like a large estate and exploded in flames.

And then, as he reached the blaze, he gave a last glance upwards—and there was Bicknell, descending gracefully just above and in front, so near that he had to turn quickly for fear of fouling his parachute lines.

Again the flood of relief was so overwhelming that for a few moments it left him limp and gasping. Then he was gliding round and round the big white canopy, leaning out and shouting wildly. Bicknell was obviously unhurt. They made rude signs at one another for the rest of the way down.

Tuck watched his leader make a safe landing—on the Southend arterial road, where he was promptly offered a lift back to Hornchurch by a pretty girl in a red sports car—a learner driver who, Bicknell related afterwards "scared me more than anything else that happened that day."

With Bicknell safely on his way home, Tuck got to grips with his own problem. His smashed wing was flapping most alarmingly now and there was no way of telling whether his undercarriage was damaged—or even if it was still there. The controls were sloppy and he knew that when it came to landing, the Gladiator was liable to whipstall at well over the normal touch-down speed. And supposing the damaged wing broke off just as he was making his approach to the airfield? There would be no time to use his parachute. It might be wiser to bail out now.

He decided to experiment a little first. He climbed her back above the clouds and, easing back the throttle, made a practice landing on a fairly flat patch of stratocumulus. He found that so long as he didn't let the speed drop below about 80 m.p.h., she gave no hint of stalling. He set course for Hornchurch.

Luckily the undercarriage proved intact, and a few minutes later he made a perfect wheel landing at base. As the ground crew dragged the damaged plane away, a hunk of the broken wing fell off and the remainder sagged to the grass. The middle-aged N.C.O. in charge whistled softly.

"Tuck's luck," he said, fingering the crumpled fabric. "Some blokes 'ave it, some don't."

5

One day in mid December, 1938, Bicknell called Tuck into his office and told him to shut the door. Bob knew this wasn't because of the cold—the flight commander's voice had an excited edge to it.

"Listen," Bicknell began the instant the lock had clicked, "you know this squadron's going to be re-equipped with Spitefires."

"Yes—*one* day," Tuck said cynically, swinging a long leg over a corner of the trestle table. They had been told this months before, but since then there had been only a series of rumors and he had resigned himself to waiting at least a year. Once or twice he had visited Eastleigh, the Supermarine Company's experimental field in Hampshire, and watched test pilot Jeffrey Quill and his colleagues putting the first production Spits—officially still on the secret list—through their paces. He could hardly believe his eyes—such speed, and such maneuverability in a *mono*plane!

Sleek, strong and incredibly high-powered, the Spit represented the fighter of his dreams, and he had fallen in love with her at first sight. Thirty feet of wicked beauty. . . . But it seemed that Air Ministry red tape and shortage of money, which made production painfully slow, would keep them apart for a long time yet.

"It's not going to be as long as you think," Bicknell said. "As a matter of fact I've got a bit of news that ought to make a damned fine Christmas present for you. Pretty soon now, Jeff Quill's going to start checking out a few chaps at Duxford. Each of the future

65

Spit squadrons will send him one pilot to begin with. And you, you lucky bastard—you've been chosen to represent 65."

Tuck just stood gaping at him. Bicknell got up from his chair, grinning broadly, and slapped Bob's shoulder.

"*I* didn't choose you, don't think that. It was the Old Man himself."

"Well, I'll be damned!" Tuck breathed, a boyish glee soaking through him—he had been so sure that his commanding officer, Squadron Leader Desmond Cooke, still considered him overconfident and irresponsible. He felt that he ought to say something rather grave and modest.

"This is wonderful. I really can't see why he should pick me—Lord knows, plenty of the chaps have more experience."

"Too much, in fact—and all on biplanes," Bicknell said, his sagacious, honest face suddenly deadly serious. "This job doesn't need experience so much as flexibility . . . a young quick mind that can accept and absorb new ideas and techniques." His mouth twisted in a tired smile. "Maybe that's just what this whole bloody air force needs—maybe we should apply for nice, new flexible minds to be issued to officers every two or three years." He paused, shrugged, and grew serious again.

"Listen, Tommy—the Spitfire isn't just another new aircraft—a bit faster and more heavily armed, and with a few more clever gladgets. The Spitfire is revolutionary, a new conception in design. At last, it seems we're catching up with progress . . . going into a new age of flying—and that's no bullshine, either!

"Which is just as well, because . . ."—he paused, turned away from Tuck and walked to the window, looking up into the dark gray sky—". . . because the bloody politicians are making a shocking mess of this German problem. Munich was a complete balls-up. My bet is before long we're going to need every Spit and every trained pilot we can find. So for God's sake don't have a crash!"

*   *   *

Jeffrey Quill was a small, lean man, just twenty-five. His puckish face seemed always to have a half smile, and with his soft casual voice and sporty clothes you might have taken him for an indolent and spoiled youth, until you noticed his eyes—brightly alert, steady and dominating. He had served in the air force from 1931 till '36, then joined Vickers as an assistant test pilot. Now—at the tailend of '38—he had been appointed senior test pilot at the Vickers (Supermarine) Works. Only one man knew more than he about the new Spitfire—its designer, R. J. Mitchell. And Mitchell, his health broken by years of financial worry and overwork, was no longer able to personally supervise the advance test program or take any part in the training of the first service pilots who were to use his brainchild as the foremost weapon of Fighter Command. Thus to the youthful Quill and his tiny staff fell the huge task, and the frightening responsibility, of getting the R.A.F. swiftly and surely mounted on this new and vastly stronger steed.

They began with half a dozen young pilots, the pick of the squadrons which Air Ministry had earmarked for re-equipment. Quill was glad when Bob Tuck turned up in this group—he instinctively liked this slim, eager young officer who had questioned him so hungrily and intelligently during their previous, brief meetings at Eastleigh. Tuck had shown that he knew by heart the few general performance facts so far released for publication. Obviously he was tremendously thrilled by the Spit and, unlike many other pilots, he was not in the least awed by her "unorthodox" appearance, nor cowed by her highlanding speed.

At Duxford, Quill personally took him out to a parked Spitfire marked K-9796 and began to teach him cockpit drill. For over an hour Tuck sat in the plane while the test pilot, peering in, ordered him to place his hands on various controls. When he had memorized the precise position of each knob, switch, lever and dial—there seemed to be at least three times as many as in the Gladiator—and could flash a hand to it unerringly, Quill made him do it with his eyes tightly closed.

Next, he had to brand into his mind the ritual letters BTFCPUR, which represented the order in which these controls were to be checked and set at the last moment before take-off—brakes, trim, flaps, contacts, petrol, undercarriage and radiator. Other, similar formulas governed the processes of climbing, descending and landing. This took the best part of another hour.

Finally he had to memorize the different speeds, temperatures, pressures, revolutions per minute and other instrument readings necessary for take-off, climbing, straight and level flight, normal cruise, fast cruise, various rates of turn, airdrome approach and landing. For a while his head was swimming, but they kept at it doggedly and soon he had mastered all this mass of vital operating data: Quill decided they'd had enough theory and it was time for his pupil to put it all into practice.

"Just one more thing," he said, as Tuck fastened his straps. "Be damned careful not to get her nose too far forward on take-off. Once the tail's up you've only a few inches clearance between the prop tips and the deck."

"Right—I'll remember."

"Fine. I'm going to leave you now. Just sit here for a minute or two and go over everything by yourself." He jerked a thumb towards a group of airmen squatting in the grass a few yards away. "When you're all set to go, give the boys a yell." Then, still wearing his lackadaisical half-smile, Quill jumped down and sauntered off.

Left alone, Tuck had momentary qualms. He felt like an impostor, sitting here in this strangely shaped machine . . . looking out through the tiny windscreen at that long, slim, shark's head cowling. . . .

It was so odd not to have the comforting bulk of an upper wing over his head and the familiar latticework of struts and wires on either side. The cockpit was so small his body seemed to fill it, and he had the curious impression that somehow he'd got himself inserted up to the shoulders in a giant bullet. And when he thought of the thousands of horsepower imprisoned in that Rolls Royce Merlin engine just in front of him, waiting to burst into thunderous action at *his* bidding—he realized this Spitfire bore practically no relation to any of the aircraft he'd flown previously. His mouth went dry and his chest was a vast, empty cavern.

There drifted into his momentarily confused brain, like a chill fog, the memory of those terrible first days at Grantham, and he saw Wills' suffering face again and he heard Wills' too casual voice coming over the speaking tube, telling him to watch the airspeed, to keep the wings straight, not to be so bloody hamhanded on the stick and not to slam the throttle about like that . . . He tightened his harness, whistling shrilly for the ground crew. They came running to the aircraft and a few seconds later he pressed the starter button. The single-bladed prop, which looked to him like an enormous plank of wood, turned slowly, hesitantly, then . . .

B-room-bang! The powerful Merlin fired with a deep, crackling roar.

"Chocks away!" Bent figures scurrying through the thin blue exhaust smoke, dragging the heavy blocks from under the wheels. . . . A dozen white needles shuddering on the dials before his face, like admonishing fingers waggling at him. . . .

Taking his time, making each movement light yet sure, he eased the throttle forward, opened the radiator wide to prevent overheating, eased off the brakes and taxied carefully to the downwind side of the field. There he turned her into the wind, stopped and began his cockpit check. He could remember every word Quill had said.

He eased the throttle open and she seemed to leap forward. At first he couldn't see ahead because his vision was obscured by that long cowling, so he had to stick his head out and keep her straight by looking along the side of the nose and using a distant tree as a reference point. From the stub exhaust, short jets of greenish-blue flame stabbed back towards his face, quivering furiously, roaring like a battery of blowlamps.

The thunder of the engine as it surged to full power beat on his head like mailed fists.

She gathered speed slowly. At just over 60 m.p.h. the tail came up, giving him a clear view ahead. Then she started to accelerate very quickly.

Don't let her nose get too far forward. . . .

Gently he checked the stick, keeping the prop tips clear of the ground. All the time his feet were working nimbly on the rudder, fighting her strong inclination to swing to the right—because at this speed the torque of air spiraling back from the prop struck the tail assembly on the right side with considerable force, trying to veer her round. Then suddenly she was in the air.

When you change hands to retract the undercart, watch your airspeed. . . .

With his left hand he throttled back to climbing revs then tightened the large, milled knob which clamped the lever there. Now, Quill had warned, came a moment of danger; the lever which operated the undercarriage was on the right side of the cockpit, so to retract, it was necessary to change hands on the stick. The retracting lever had to be moved with the right hand to select "up," and then pumped vigorously for several seconds. It wasn't at all easy to perform large, energetic movements with one hand and at the same time keep the other completely still. At this low altitude, before the new pilot had had a chance to get the "feel" of the aircraft, a comparatively small, jerky movement on the stick with the left hand could cause either a stall or a dive.

Even the best of beginners, Quill had added indulgently, couldn't be expected to hold her in a steady climb at this point—at least a gentle "porpoising" movement was inevitable.

He moved his hands unhurriedly. Resisting the temptation to give just one quick glance sideways and downwards towards the undercarriage lever, he kept his eyes on strict vigil, flicking between the airspeed indicator and the distant, hazy line of the horizon. The nose remained precisely in the correct climbing attitude while his right hand sought and found the lever, moved it upwards, pumped it steadily.

The Spitfire's hydraulic muscles flexed, swiftly and smoothly drawing up the undercarriage. *C-loomp c-loomp*—the genteel thuds under his feet as the wheels folded into their belly housings suggested sheer luxury, like the cushioned slammings of giant refrigerator doors. A current of elation sparked through him. He could feel in his hands the touch of a sure and responsive mechanism, the delicate poise of this machine, light and quick and graceful as a ballerina. As he slid the hood closed he chuckled aloud. Here in this transparent, streamlined carapace, looking out along the long, tapering nose, he felt drunk with power and speed.

Above the winter clouds, for twenty minutes or so he frolicked joyously—stalling, spinning, looping, rolling. He found himself thinking of this machine as a live creature—gentle and sensitive, with a great heart that throbbed bravely. An understanding and intelligent creature that responded instantly to the most delicate suggestive pressures of its master's hands and feet. He had never dreamed that flying could be like this: he knew that with a little time he could make this plane almost a part of him—like an extension of his own body, brain and nervous system.

Descending in a long, gentle dive, he touched 360 m.p.h. Leveling out near the airdrome, he had to force himself into a more sober frame of mind in order to concentrate on the difficult job of landing a new aircraft for the first time.

He throttled back, opened the radiator wide, slid back his hood and began his approach. At first everything went well. As he turned across wind, he lowered the undercarriage and flaps. Next, turning into wind, he put the prop into fine pitch and eased the stick back: then the long nose reared up and blotted out the ground ahead!

For the barest second he was thrown off balance. And then he remembered. . . .

On the last part of your approach it's a good idea to

*put her into a gentle sideslip, and then straighten her
out just before the wheels touch. . . .*

A little rudder and a few degrees of opposite bank,
and the nose obligingly swung a little to one side.
Now the Spitfire was moving crablike through the air,
forward and downward in a sort of controlled fall. The
wind shrieked in through the open side of the canopy,
full in his face, as he looked down at the tilted grassland hurtling to meet him at unnerving speed.

Twenty feet, ten feet, five . . . *now!* He centralized the
rudder and took off the bank. Like a soldier coming
smartly to attention, the aircraft snapped into the landing attitude.

Stick back, gently but firmly back into the pit of the
stomach. A soft jolt, and she was rolling bumpily over
the grass. Keep her straight! A touch on one brake,
then the other, and she was slowing . . . slowing . . .
coming to a standstill in the center of the field.

How quiet and still was the world! He relaxed for a
moment and then, wrapped in a deep, warm gladness,
taxied leisurely to dispersal. He knew with a startling
certainty that in these last few minutes he'd reached
an important milestone in his life.

He could not know *how* important.

*   *   *

He stayed at Duxford for over a week and returned
to Hornchurch on January 9th, 1939, as one of the air
force's first qualified Spitfire pilots. The following days
were one constant, exhaustive interrogation, but he
never wearied of describing to his eager colleagues
how wonderful the new fighter was.

Now the Gladiator seemed like an amiable, somewhat rheumatic old goose, and he seethed with impatience as weeks went by and there was no news
about re-equipment. Not until the end of March was
he ordered to Eastleigh to collect the first of the new
aircraft.

In the mess at Eastleigh he found Jeff Quill and several other pilots he knew. Everyone was in high spirits,
and the half cans of bitter rose and fell merrily. They

talked, of course, about the Spitfire, and all the future wonders promised by its success. They talked, too, about another new fighter, the Hawker Hurricane, which somehow had received a great deal more publicity than the Spit, although undoubtedly it was slower, heavier and not nearly so agile.

The Hurricane had smashed a number of records on cross-country flights, and the pilots of 111 Squadron—the only unit so far equipped with this airplane—were shooting terrific lines about how much skill and guts it took to handle it. They inferred that it was a kind of monster that had to be tamed and constantly watched—a machine with a sly, malicious intelligence of its own, a killer plane that only the dashing, devil-may-care elite were permitted to fly.

Undoubtedly 111 had started all this as a huge joke, but the joke was being taken seriously by the public

and the press, and the Air Ministry hadn't lifted a
finger to dispel the myth. Hence Hurricane—not Spitfire—was the name on every schoolboy's lips and in
every newspaper article on Britain's air defenses: anyone who flew it was regarded as an "ace" by the
awed, and misinformed, populace.

(This state of affairs must have been galling for the
Supermarine Company and for the squadrons taking
over the new Spits, but Tuck maintains that it had a
more serious repercussion too: for some years most
newly trained pilots held the Hurri in almost supersti-
tious dread, and without doubt a number of them
crashed while learning to fly it through lack of confidence or sheer nervousness. Their fears were quite
groundless, for the Hurricane was in truth an infinitely
more sedate and tractable machine than the supersensi-
tive Spitfire, which reacted fullbloodedly to the slightest
movement of the controls—right *or* wrong. Had official
steps been taken in 1939 to strangle the legend of the
"killer plane," after the outbreak of war he believes
there would have been fewer serious crashes among
trainee pilots of Fighter Command.)

Then they talked about another new monoplane
fighter, the Messerschmitt 109. There was not much to
say for none of them knew a great deal about it. All
Tuck could say on the subject was that he'd bet his
life it had a jolly fine reflector gunsight! Everyone
roared with laughter, for they'd all heard the story of
Milch and Udet visiting Hornchurch. Their laughter
was still ringing in his ears as he left, gay and excited
as a bridegroom, to collect his Spit and carry her
with loving hands across the threshold of 65's
home.

He started taxiing toward the downwind end of the
field so fast that once or twice the tail kicked up and
the machine seesawed clumsily. Then he saw Jeff
Quill's car racing across the tarmac to intercept him,
and eased up a bit.

Quill was waiting for him at the take-off point,
smiling as ever. As the plane rolled to a stop, facing into wind, the test pilot mounted the wing at a single

agile spring and said: "Better have a quick little brush-up, eh?"

Tuck had trouble concealing his feverish impatience. He didn't need a refresher—every detail of Spitfire procedure was ready now in the forecourt of his mind, burnished bright by daily polishing—but, after all, it was the best part of three months since he had flown a Spit and he realized that Quill had to be sure. So for the next five minutes he sat there, docilely practicing the cockpit drill he could have done in his sleep, moving his hands over levers and switches familiar to him as the pockets and buttons of his uniform.

"Check your magnetos," Quill said at length, getting down from the wing.

Bob locked his brakes, full on, and advanced the throttle. The Merlin's stunning roar seemed to fill the world. The whole aircraft juddered, straining to be off. And behind, the boundary hedge seethed and lashed under the slipstream's torture.

He flicked first one magneto switch, then the other. Carefully he made sure that there was no big drop in the number of revolutions per minute when she was running on only one mag. Finally, he opened up to almost full power, to ensure climbing and take-off revs. Then he throttled back and gave Quill the thumbs up sign to show him all was well. Quill waved cheerfully, signaling him to take off. The Merlin's roar rose to a crescendo and in what seemed less than a minute he was streaking up through wispy clouds and Eastleigh was a diminishing patch on the cluttered landscape far below.

High over Hampshire and Essex, riding home in his new Spitfire, he hummed a theme from Mozart, and with the canopy closed the deep drone of the cruising engine seemed to form great chords of harmony. Now and then he plunged into a puff of cloud, pulled up out of it again in a steep climb—playful as a young seal in a summer sea.

He had never been so content. A feeling of inner peace and welling strength filled him, but he entirely failed to recognize the truth: that in some weird way this machine—this marvel in metal which he understood so well, and which seemed to understand him—already was at work altering his character, completing his growing up. He only knew, vaguely, that suddenly he had lost some of the old restlessness, and found an inner balance.

6

It was a few minutes before 11 a.m., on May 23rd, 1940, when he saw the shooting war for the first time. From 15,000 feet he saw the immense, black pall of smoke brooding over the Dunkirk beaches, the sullen flashes of the enemy guns inland, the occasional puffs of flak staining the morning sky, and—quite distinctly—the straggling columns of men, threading their way painfully through narrow, choked roads, down to the sea.

He saw several little ships, privately owned motor launches and trip-round-the-bay pleasure boats from Brighton and Cromer and Broadstairs and Cowes; with their garish paint, somehow they seemed in the worst of taste now, like aging actresses emerging from the playhouse into the world of grim reality still wearing their stage make-up. Yet they plodded gamely to and fro among the bigger, sleeker vessels, the gray-uniformed soldier-ships that were too few for the huge task ahead.

He saw shells exploding on the cluttered beach and raising tall, white plumes in the shallow water. Smoldering trucks and carriers, ruptured convoys, abandoned stores piled ready for burning, gun pits being dug in the dunes, sandbags being filled for the last dour stand here on this broad, white strand which, only nine months before, had known the light touch of sandal and beach ball and the barefoot scamperings of children.

All this he saw, and curiously, could not understand how remote it seemed. He might have been watching a newsreel or studying newspaper photographs. He knew his brother Jack must be somewhere down there, fighting with the 78th Field Regiment, Royal Artillery, but even this extremely personal link with the chaos below awakened no bitterness or anger. He felt exactly as he would have felt if at that moment he had been flying a practice formation over Shropshire, or one of those pointless patrols over the empty North Sea which he had carried out so often during the last few months.

A flight of Hurris slanted past far below, heading home to refuel and rearm. He wondered if Caesar Hull was one of them. He envied Caesar, for the tough little Rhodesian had fought in Norway and bagged five Huns. He envied *all* Hurricane pilots, for Hurris had been scrapping in France as well as Norway while the Spit boys were kept patrolling the backstreets of the air war—because Air Chief Marshal Sir Hugh (later Lord) Dowding, the C.-in-C. Fighter Command, had decided to hold them for Home Defense.

Up till today the war had been for the most part boring, exasperating for him. A regular officer with nearly 700 hours in his logbook and "exceptional" gradings in every branch of flying, and he'd had to sit at home while youngsters of considerably less experience went out to mix it with the Luftwaffe. . . . On the Spit squadrons tempers had flared. Pilots were frequently sent on wild goose chases, misdirected by inexperienced ground controllers and fired on by excitable A.A. batteries on the south and east coasts.

But on May 1st he had been posted from 65 to 92 Squadron, then based at Croydon, where the C.O., Squadron Leader Roger Bushell, had immediately given him command of a flight and promoted him to Flight Lieutenant.

"We'll be getting a crack at them soon, don't worry," Bushell had said in his deep, refined voice. "You're just the sort of bloke I need, and I'm bloody well going to work you till you're on your knees. We've got to lick this squadron into top shape in double-quick time. Go and dump your kit. I'll meet you at dispersal in ten minutes."

Ninety-two had been formed on the outbreak of war from a nucleus of Royal Auxiliary Air Force pilots—businessmen and students who in peacetime had given up their weekends to fly, some of them at considerable personal expense. The remainder were Volunteer Reserve pilots, straight from the flying schools. They had only recently been given Spitfires Mark II. In theory, a fighter squadron's full complement at this stage was twenty-six pilots, but in fact 92 had just fourteen or fifteen.

Tuck had expected to find them very different from the "professional" pilots of 65, and was astounded and delighted to discover the atmosphere of the Croydon mess even wilder and woollier than Hornchurch. Moreover, in some mysterious way these wartime chaps had managed to learn the entire range of R.A.F. slang and idiom, all the rude songs, all the traditional nicknames ("Chippie" Wood, "Shady" Lane, "Dickie" Bird, etc.,) and, most surprisingly, they had the same cynical approach to anything that savored even faintly of militarism, rhetoric or red tape. Even the youngest had acquired the languid arrogance of experienced pilots.

Because the Spit was new to most of them, and because each one realized the "phony war" was over and that very soon—perhaps within days, or hours—the squadron would go into battle, they had practiced combat drill with a quiet ardor. Between flights they had squatted in a circle on the grass and listened to Tuck, wide-eyed and attentive as children listening to a bedtime story.

They were a cosmopolitan bunch. There were two Canadians, "Eddie," Edwards and John Bryson, a former member of the "Mounties;" Howard Hill, from New Zealand; Pat Learmond from Ireland, and Paddy Green from South Africa. For administrative purposes the squadron at this time was subdivided into two flights of six or seven, but in the air they flew in three subunits of four. Tuck's flying section were all English—Bob Holland, Allan Wright and Sergeant "Titch" Havercroft—so short that he had to have two rubber cushions under his parachute before he could see over the instrument panel. John Gillies (son of the distinguished plastic surgeon), Peter Cazenove, Roy Mottram, Hargreaves, Bill Williams and Tony Bartley completed the "native" contingent. Green led the second flying section, and the C.O. headed the third. The two or three remaining aircraft, with their pilots, were held as reserve.

At this early stage of the war most fighter pilots still wore the white flying overalls which had been part of peacetime bullshine. Their leather helmets had built-in earphones, and a rubber attachment combining oxygen mask and microphone was clipped across the lower half of the face like a visor. With the new large, tinted goggles in place and the visor clipped on, the face was almost completely blanked out, and the general effect was decidedly sinister—robotlike.

The broad, tough canvas straps of the parachute harness ran over each shoulder, down the back, through between the legs and up over the belly. All four ends clipped into a strong spring lock in the vicinity of the navel. This lock had a quick-release device: after the 'chute had landed him safely on land or in the sea, the pilot had only to twist a circular metal plate on top of the lock and then rap it smartly with his palm or clenched fist to be instantly freed from the harness, the tangled lines and the drag of the billowing silk. There had been a few strange cases in France of pilots operating this device while still in midair, and plummeting to their deaths. The only possible explanation seemed to be that those men had been suffering great agony from wounds or burns, and released themselves deliberately.

The parachute pack was worn like a huge bustle, but very lowslung, so that it fitted into a recess in the bucket seat when the pilot sat down in his cockpit. A few weeks after Dunkirk a smaller, very ingeniously constructed pack was added, also serving as a cushion, which contained a rubber dinghy that could be swiftly inflated by compressed air bottles attached to it. The same inflation device was incorporated in the "Mae West," a bright yellow, rubber lifebelt which was worn like a waistcoat under the parachute harness. (Later versions of both dinghy and "Mae West" were constructed to inflate automatically once immersed, so that injured or unconscious pilots could remain afloat.)

Following the example of Hurricane squadrons in France, the majority of the 92 pilots carried revolvers —usually not in holsters, but stuck down the legs of their heavy flying boots. If a man landed in enemy territory there might be a slim chance of shooting his way out of trouble and getting through to the British lines, but—more important—a revolver provided the surest and quickest way of destroying a crash-landed fighter. A few bullets pumped into the fuel tank produced a consuming inferno.

Bushell, was a South African who had practiced in London as a criminal lawyer before the war. He was better known as a champion skier who had coached the British Olympic Team.

Often he had a single tear coursing down his left cheek, the result of an eye injury received in a skiing accident a few years before. No one had ever seen him wipe the tear away—he simply ignored it, even when the eye streamed for hours on end. His sight was unimpaired, and this inconvenient ailment had only one result: he always held his head slightly tilted to the right.

It was hard to tell how old he was, though he must have been about thirty. Of medium height, dark and indelibly tanned, he was wonderfully compact and fit. At practice "scrambles," when the pilots sprinted across the grass, vaulted into the cockpits and strained every fiber to get into the air in a matter of seconds, the "old man" nearly always won.

Bushell's voice was strong and cultured, and his diction was like an actor's, yet there was no false, courtroom pomp about him and he could bellow cheerful vulgarities and roar the bawdiest choruses with the rest of them. He was never one to stand on ceremony, but a candid and open character, ready to joke or play the fool with all. Yet when there was serious talk, the affability and the humor fell away and he became direct, shrewd, commanding. He engendered a fine squadron spirit with his overwhelming, enveloping aura of personality and strength.

On the morning of the 23rd they had moved to a forward base, nearer the coast—Tuck's old home, Hornchurch. There they breakfasted, refueled, got briefed. And now at last, after all the waiting and working, the patrols and the play acting, the testing time had come—and here they were in combat formation with Bushell leading them over the blazing beaches and their gunbuttons set to "fire," with their faces rigid and aching with the suspense and their necks in swivel gear and their eyes smarting from scouring the furthermost corners of the sky for the

first tiny glint of metal. Twelve Spitfires, marching proudly, challengingly, disdaining the cover of the stratocumulus cloud 5,000 feet higher, inviting the enemy to attack.

But as they flew down the coast on their specified "offensive sortie"—Calais—Boulogne—Dunkirk—nothing rose to meet them but the occasional flak-puffs and the oily smoke—though they had been warned at briefing to expect formations of up to forty Me. 109s, and had been given the impression that long, un-broken lines of Ju. 87B Stukas were stretched across France, lining up to dive bomb the cornered British troops.

They flew through a patch of turbulent air and their wings rocked a little, the whole formation buckling and writhing momentarily out of its purposeful shape. It was all over in four or five seconds, but it left Tuck very uneasy. The formation was too tight—he couldn't see any point in keeping so close together when at any moment they might have to tear their minds and eyes away from each other to concentrate on battle. . . .

"Turning right—go!" Bushell's voice sounded gruff over the r/t. He was probably just as angry as ev-eryone else.

The twelve Spits banked in one even, coordinated movement, directly above Dunkirk (2) town, and be-gan wheeling through 180 degrees to go back the way they had come. They would fly up and down this coast until either they were attacked or their fuel tanks were nearly empty and they were forced to return to base. Tuck told himself that somewhere along this stretch, between here and Calais, they must surely meet their enemy. And yet still he felt no apprehensive tightening of muscles or nerves. He just sat there, not really be-lieving all this, observing every detail of the spectacle around him with mild amusement and a begrudging, slightly suspicious curiosity—as one inspects the scenery of a dream. He was, for instance, struck by the sheer, majestic beauty of the wheeling echelon of Spitfires, stepped up so neat and steep on the outside of the turn, each machine printed sharply on the gray-blue canvas of space.

The two aircraft closest to him were flown by Hol-land and Wright. By turning his head, he could look straight into either man's eyes, for at this moment they were keeping their positions in the turn by watching Tuck's wingtip. They seemed so close that he felt he could stretch out a hand and touch them; yet he knew that in fact they were all three infinitely remote from each other, that if one of their engines stopped or a shell set one of the machines ablaze, there was no way for the others to help and they might as well be a million miles away.

He shook himself free of the thought and as the formation completed the turn he gave Holland and Wright a cheerful wave. They acknowledged with vigorous, pumping V-signs.

Soon after that a short, strangled cry in his earphones jerked him forward against his harness, breath congealed, eyes straining. He had no idea who had given that cry, and there was no chance to find out, for the very next second, glancing downwards over his left shoulder, he saw Messerschmitts shearing in.

"Here they come—eight o'clock!" he yelled. And then Pat Learmond died. An eye searing, loud explosion, and his machine simply was not there any more. There was only a big, pulsating ball of flame, which seemed to keep position in the echelon and fly on for several seconds before it sank away under their wings. And during these seconds, while the Spit pilots were twisting in their seats to look round and see their attackers and where they were coming from, the orderly, businesslike formation began to buckle, stretch, break up. Bushell was rapping out orders, but now everybody was yelling at once and his words were lost in the babble.

The 109s were dropping out of the stratocumulus behind and slightly to the left, a long, straggly line of them, diving hard and building up terrific speed before they swept in from just under the tails. Tracer was lashing the Spitfires, chivvying pieces out of them. The leader—it must have been this one that got Learmond—came snarling clean through the heaving, splitting formation, then climbed hard for the cloud ahead. Bob decided to go after him, though he knew he had very little chance of catching up.

Climbing at full power, he couldn't match the high angle of ascent which the Messerschmitt had earned with his earlier dive. He was still a good 2,000 feet under the cloud when the German vanished into its woolly sanctuary. But he kept on, up through the glary white stratocumulus, not quite knowing what he hoped for now.

The cloud was a lot thinner than he'd expected. After only two or three hundred feet of it he broke out into peerless blue. The foe was dead ahead, not more than 2,000 yards distant, racing eastwards for his base, skimming so low over the great slab of milky cloud that he blended with his own shadow, like a flatfish scuttling over sea-rippled sand.

Tuck realized that his shallower angle of climb had actually brought him closer to his quarry. The German was flying perfectly straight, so apparently he felt quite safe up here. He might even be cruising at normal revs, in which case Tuck could overhaul him by using full "emergency" power.

He pushed the throttle forward hard—breaking through the light seals that guarded the "emergency" slot—and settled the Spitfire as low as possible, cutting through occasional wisps and peaks that thrust up from the unbroken cloud; sometimes, in places where the surface was smooth, he flew for seconds on end with only his head and the canopy in the clear. The Messerschmitt kept straight on, and very gradually it grew larger in his windscreen.

He hasn't seen me, he hasn't seen me. . . !

For the first time since take-off, a full-blooded thrill flushed through him, but immediately he suppressed it: this was the time to be cool, relaxed, methodical, precise . . . he must never let himself get excited or emotional in battle, never! To keep his place as an exceptional pilot, he must prove himself as level-headed and as skillful in combat as in all the other flying he had done thus far. He *had* to be exceptional—ever since Grantham there had been no other way of facing each day. If he were not exceptional, he would be nothing.

Grantham, Hornchurch, Duxford, Eastleigh, Northolt —all one long rehearsal. Now the curtain was rising —this was the moment he had been working for, training all these years. He must not fumble. For him failure was out of the question. It was different for all the others on the squadron, volunteers and auxiliaries—this was his chosen profession in life!

The Messerschmitt made a gentle turn of about ten degrees to the right, a small course correction, and he eased round after it. As it banked, its long ganoid body and the sharp snout, slicing through a knoll of foamy cloud, reminded him with jolting impact of the sharks he had shot from the rail of the *Marconi*. He reached out and adjusted the reflector sight for the wingspan: Me. 109.

Sixteen hundred yards . . . fifteen hundred. . . . He was gaining steadily, and the Hun was still unaware of his presence. He could see the Luftwaffe markings very plainly now. There they were, big black crosses—exactly as he had been told, exactly as he had seen in photographs and aircraft recognition posters. For some curious reason he experienced a twinge of disappointment, as if he'd been hoping for something different, *un*familiar.

The enemy's silhouette continued to expand. He'd been chasing him for over two minutes now, and they must be well inland. He kept glancing in his rear vision mirror, and sweeping the sky all around. The vast blue vault was empty except for the two of them. He checked his fuel, oil and radiator temperatures, all in one swift, custom-learned sweep of the eyes. All was well, everything was ready; each tiny piece of mechanism was toiling nobly, performing its own intricate function. The Spitfire would not blunder—only he, the human element, could wreck the whole elaborate scheme by a single wrong decision.

He didn't give a thought to the other human element —the man in the Messerschmitt, the man who was going to die for the cardinal sin of not looking over his shoulder. There was nothing personal about this.

Twelve hundred yards . . . eleven hundred. . . . Now the Messerschmitt's wingspan almost filled the space between the parallel lines of the reflector sight. Another few seconds would bring him up to maximum range.

All at once he noticed a broad, black highway running almost parallel with them across the white desert of cloud. It nagged at his attention, dragging his eyes away from the gunsight, puzzling him with its startling

evenness and straightness. Then he caught the faintest
scent of burning oil, and knew what the dark stripe
was: the eastward-drifting stratocumulus was collecting
the smoke that rose in a great pillar from the Dunkirk
beaches and dragging it, like a black ribbon of mourn-
ing, slowly inland over France.

One thousand yards—maximum range!

The 109's wingtips touched the sight's glowing reti-
cles. He put his thumb gently on the gun button and
willed himself to relax, to let every muscle liquefy, as
his father had taught him back in the days when the
school rifle team was training for Bisley. Get comfortable, hold your breath, keep perfectly still—even
a man's own heartbeats could upset his aim at the vital
instant when he exerted firing pressure!

He flicked a last glance at the turn-and-bank indicator. The needle was in the center, perfectly still, as
though painted on the dial. He took a deep breath.

And then he changed his mind.

Why open up at extreme range if he could get
closer? At a thousand yards even perfect shooting
might only damage the Messerschmitt. Every yard
nearer increased the chances of complete destruction.
Provided, of course, that the German didn't look
round. . . .

Nine hundred yards, seven hundred, six hundred.
. . . It seemed impossible that he could get this close
without being seen—in fact, he was beginning to feel
that by now the foe must *hear* him. But the 109 was
still sitting up there flying unconcernedly down the
mild May sunlight.

Five hundred.

A deep breath . . . don't tense up, stay relaxed . . .
turn-and-bank needle centered . . . keep the red dot
right on his canopy and, gently, gently, *s-q-u-ee-ze.* . . .

The eight Vickers .303 burst into life with a sound
like the crash of a roller-coaster. The Spitfire quivered
lightly in his hands. He could see the wicked little blue
flashes his bullets made as they struck the 109's wings
and canopy. For two . . . three . . . four seconds they
hosed into it, and nothing was happening. The German flew on, straight and level, with the little blue

flashes all over him like lights on a Christmas tree. Then the machine seemed to flinch, and give a nervous shudder. Its nose came up lazily and it climbed for a little, with the Spitfire following. Suddenly there was a puff of dark blue smoke and a few small pieces of metal came whipping back at the pursuer, spinning like boomerangs, as though in a final, hopeless gesture of defiance.

Tuck took his thumb from the button and ruddered smartly to the right as the burning Messerschmitt rose steeply, slowed, and seemed to hang in the air, nose pointing to the sun. Then suddenly it flicked over to the left. Once, twice, three times it rolled, as correctly as if giving an airobatics display, and then it started down in a weary spiral.

He dived through the cloud and waited underneath a long time before it appeared. Though it was against orders, he meant to follow it down, to see it crash— *he had to be absolutely certain.* But it was hard to follow: several times as it built up speed its elevators, trimmed for level flight, made it climb again for a few hundred feet, so that it came down in long, dipping stages. It was as if the Messerschmitt were trying to shake him off, to sneak away and die alone.

But finally it fell, in its red and black garland, into a plowed field. And blew up.

He climbed back into the cloud and set course for home. He felt no elation, only a quiet satisfaction. He had carried out his first act of war—coolly, correctly, professionally.

Not even Wills nor Savile could have found fault!

When he landed at base he found most of the squadron already refueled and rearmed, standing by for another "scramble."

Paddy Green, John Bryson and Tony Bartley each had bagged a 109. That made the score five for the loss of poor Learmond, whose incandescent wreck had been seen to crash smack on the French beach. Bushell was kicking at the ground in disappointment. He'd followed a Hun for three minutes, then lost him in cloud before he could get a squirt at him.

They gulped tea and munched sandwiches, excitedly comparing their experiences. A van arrived with five of the new bulletproof windscreens, completed at the factory that morning. The ground crews worked with demonic energy, and Tuck's machine was one of the three they managed to refit.

The thick, nonsplintering perspex was the only protection Air Ministry could offer: at this stage the Spitfire was entirely without armored plating.

They were off again at 1:45. One of their two reserve pilots had taken Learmond's place in Green's flight, but otherwise the formation was the same. At 8,000 feet over the French beaches several of them simultaneously spotted a pack of between thirty and forty twin-engined Messerschmitt 110s diving from almost directly overhead. Bushell immediately put the squadron into a turn to the left, forming a sort of defensive circle in which they could, at least to some extent, protect each other's tails. Tuck, again in that strangely cool, almost dreamy state of mind, couldn't help thinking of Western movies he'd seen, in which covered wagons adopted the same tactics to fight off redskins. . . .

The 110s were formidably armed. In the nose they had a fixed cannon and machine guns, fired by the pilot. Behind the pilot, back to back with him, the rear-gunner manipulated a single free-swiveling, heavy machine gun. They came through the formation spewing tracer fore and aft, and then suddenly friend and foe were milling round and round. Now some of the Spitfires were opening up. The air was crisscrossed with their fire—a fantastic imbroglio of glowing, curving lines—and clearly there was a great risk of pilots on both sides being shot down by their comrades.

Wheeling round in the thick of this confusion, for a time Tuck could see nothing to shoot at. Then a 110 drifted majestically in front of him, startlingly close, in a gentle bank and skidding slightly. It seemed enormous, impregnable. Its rear gunner swung his long barrel round and Tuck instinctively ducked as the bullets ricocheted off his cowling and canopy, filling the cockpit with the bitter stench of cordite. Then he was firing back, holding the gunsight's dot steady on the gunner's goggled face for seconds on end, watching the Spit's eight streams of metal tear into the Messerschmitt's cabin, fuselage, and port engine.

At last the tail-gunner stopped firing. Tuck thought by now the fellow must be carrying more than his own weight in bullets. The 110 was seesawing, bits were flying off the port wing, and a thin, straight line of black was trailing from the port engine. He kept his thumb on the button until it flicked over and plunged vertically down, blossoming like a red flower as flames burst from its shattered engine and spread along the wing.

He pulled up and put his neck into swivel gear. The sky about him was a streaked and mottled hiatus. Spitfires and 110s were slashing across and up and down, turning, writhing, rolling through the lethal latticework of tracer. His earphones rang with a bewildering din— four and five voices shrieking at once.

"Look out—here's another!"

"Watch that bastard—smack underneath you, man, *under* you!"

"He's burning—I got him, I *got* him chaps!"

"Bloody hell, I've been hit. . . ."

"Jesus, where are they all coming from?"

"For God's sake, *some*body. . . !"

This was all wrong—the r/t shouldn't be abused like this. Properly used, it ought to be a valuable life preserver, the teambrain of the squadron. Why didn't they keep quiet until it was absolutely necessary to shout a warning? Why didn't they use callsigns, or even Christian names, and give heights and directions as they had been taught—"Two coming in four o'clock, above you. . . ." In their excitement they'd abandoned all set procedure and resorted to meaningless, unprofessional clamor.

Infuriated, he yelled into his microphone: "Shut up, for Christ's sake!" But it did no good.

Shadows flicked by close overhead—a 110 with Tony Bartley not more than fifty yards from its tail, pouring continuous fire into it. Bartley was taking a terrible chance: if the Messerschmitt exploded the Spit would be destroyed too. "Bloody idiot!" Tuck muttered—but the term expressed affectionate admiration.

Then above the commingled dins there came a cry that almost split his helmet—a long, horrendous sound which might have been dragged up from secret ocean depths. To his right he saw Flight Sergeant Wooder's Spitfire a blazing brand. Under the perspex of its canopy there was not the vaguest sign of a human figure, only a seething mass of yellow flame. It was like looking through the peep hole of a blast furnace.

The fiery wreck fell slowly, wilting, becoming shapeless, till it resembled nothing so much as a lump of rag which somebody had soaked in gas, lit, and tossed over the side. Tuck watched in sickly fascination—until something struck his windscreen a tremendous, jarring blow. A 110 was charging towards him, head-on, bombarding him with its shells and bullets.

He blazed back at it, holding the collision course. The two machines hurtled to meet each other, converging at a speed of roughly 600 m.p.h., their gunports flashing like bared fangs. The first to break off would almost certainly be destroyed—because he would present an easy target to the other. It was a question of which pilot had the strongest nerve.

In the final second, when it seemed that the weird joust could only end in an annihilating crash, an irresistible reflex made Tuck close his eyes and yank his head down below the level of the windscreen. A split second roar, like an express train bursting out of a tunnel. He never knew whether the Messerschmitt passed below or above him. When he raised his head again, and looked over his shoulder, he saw it far behind him, losing a little height, turning east—towards the land. He thought he glimpsed a faint, white trail from one of its engines. He wheeled the Spitfire and set off in pursuit.

On the top left and bottom right corners of his windscreen there were ugly, opaque blotches, staring in at him like big blind eyes. Two of the Hun's shells had smacked him squarely: if the fitters hadn't installed the new type screen just over an hour before, these hits would have come through and taken off his head. . . .

He caught up with the 110 a mile or two inland and

opened fire from about 500 yards. The rear-gunner replied with great accuracy, and the bullets drummed on the Spitfire's nose and canopy like savage rain. The German pilot took violent evasive action, lifting and dropping his damaged machine in a mad frenzy. But he couldn't shake off his resolute and agile tormentor, so he dived to treetop level in hope the Englishman wouldn't have the nerve to follow.

Tuck followed, hard and fast. Now the duel became an obstacle race, with each protagonist's attention divided between fighting and flying, with neglect or misjudgment in either department entailing equal peril. They squeezed sideways—banked and skidding— through narrow gaps in thickets of firs . . . lifted wing-tips over church steeples . . . grazed farmhouse chimneys with their prop tips . . . weaved along a river so low that their slipstreams whipped up spray and set small boats rocking madly at their moorings. And all the time they were exchanging long bursts of fire, knocking pieces off each other, the pursuer drawing closer with each second until they were less than two hundred yards apart.

Once Tuck came near to losing the battle. The 110 went under some high tension cables and barely managed to pull up in time to clear some sharply rising ground beyond. Bob, on a split-second decision, pulled up earlier, rose over the cables like a hurdler—and in so doing exposed his belly as a succulent target for the Me. rear-gunner. In the dragging moments before he could get flat down on to the deck again, scores of bullets ripped through his wings and the underside of the fuselage. But the Merlin's warlike roar didn't falter and the Spit flew on, cool and competent and unperturbed.

It was the rear-gunner's last gunfire. Tuck's next burst riddled him; the long-barreled gun dipped, swung loosely, then drooped dispiritedly over the side of the Messerschmitt's fuselage. At once the German pilot, realizing he was at the Spitfire's mercy, swung into a long, flat field and crash landed in a whorl of dust.

Tuck circled about fifty feet over the wreck and saw the pilot scramble out and run clear. Jolly good luck to him!—he deserved to get away with it, after that wonderful bit of low flying!

No hard feelings. Flushed with success, excited from the long and hazardous chase, Tuck felt a positive affection for his vanquished foe. So he circled even lower, slid back his canopy and waved to him.

The German was standing perfectly still now, a few yards from the hulk of his aircraft, looking up at the wheeling conqueror. An erect, proud figure in a smart gray uniform and polished black boots. And then suddenly he raised his arms.

He's waving back. No hard feelings. . . .

A bullet whined through the cockpit, chipping the edge of the windscreen's side panel, less than six inches from Tuck's face. Another glanced off the outside of the screen—leaving a third opaque blotch. The German's raised arms held a Schmeisser automatic machine pistol.

Tuck felt betrayed. So this was the enemy—this was the sort of fight it was going to be. . . . His eyes narrowed and his teeth clamped hard.

He drew away from the field, then went in again, very low, and laid his sights on that defiant gray figure which still held the gun leveled at him. Coolly, carefully, he checked his turn-and-bank indicator, then squeezed the button. The ground all round the German erupted in wicked flashes, clods of earth and spurts of dust. It was a very short burst, for the Spitfire had very little ammo left, and after a second or so there was only the hiss of compressed air and the futile clankings of the empty breech blocks. Enough, though.

The Hun staggered two or three steps through the dust, like a man lost in a fog, then pitched on to his face. The oily smoke from his burning plane gradually drifted over and covered him.

*   *   *

Tuck joined up with another Spitfire midway over the Channel. It was Tony Bartley, who greeted him flippantly.

"I say, old boy, what you got there—a piece of lace?"

"You've collected a few holes yourself," Tuck told him. "Better stick your arms out and start flapping."

Each made a careful survey of the other's aircraft and exchanged details of visible damage. Then they arranged a bet—on which of them had the most holes. The loser would stand the winner free beer all that evening.

As they crossed the south coast Tuck's engine began to rasp and clank and falter. The cooling system was useless, and rising pressures and temperatures were bending the dial needles against the stop-studs. He throttled back a little and nursed her along, losing precious height but keeping the prop turning.

Tony opened his canopy, shut off his engine and glided in close, listening. "You'd better go in first, old boy," he offered. "You sound like a tinker's barrow." Tuck accepted, gratefully.

On his approach to Hornchurch the engine finally seized up and, robbed of power, he could only land her on the grass verge in front of the control tower. As he got out a gray Humber squealed to a stop nearby and the station commander, Group Captain "Daddy" Bouchier, came striding over, booming angrily.

"You can't leave that aircraft here, man! Damnitall, don't you know the drill? Get it over to dispersal immediately—this area must be kept . . ." His voice trailed off as he spotted the black rash of bullet and shell holes which covered the whole aircraft. He walked right round it once, slowly, in silence. Then he said, "Yes . . . yes . . . you *are* in a bit of a mess, I see."

They faced each other across the port wing, which was punctured in half a dozen places, and suddenly, simultaneously, they began to grin . . . to chuckle . . . to roar with laughter. For each of them—for the white-templed commander, the victor of old air battles who realized the long, high fury which had burst upon this small, strained force; for the young pilot, keyed up and breathless and sweating from his first day's combat —laughter was purely a nervous reaction. Nature had taken over after the hours of tension, commanding them to relax for a moment: the laughter unburdened, refreshed, strengthened.

But half an hour later, when the squadron's losses for the day were known, the whole station was grim and very quiet. Flight Sergeant Wooder . . . John Gillies . . . Pat Learmond . . . and the C.O., all shot down. Paddy Green had stopped a hunk of armorpiercing shrapnel in his thigh. He'd flown back, faint and vomiting, with his thumb stuck in the wound, pressing on the smashed arteries. The M.O. was satisfied that the leg could be saved, but he would be laid up for a long time.

True, 92 had bagged at least twenty Huns in the two sorties, but the fact that stunned everyone was that in a single day's fighting nearly half the squadron had gone.

Roger Bushell had last been seen chasing a 110 inland at low altitude. A second German had "bounced" him from above, cutting him off from the coast, and scored several hits with its first long burst. It looked like "the chopper."

Bouchier came into the mess about eight and took Tuck aside.

"Tucky, I'm giving you the squadron. How d'you feel about it?"

Tuck fought down an uneasy stirring in his lower chest, and felt it give way to a warm glow.

"All right, sir. Thank you."

"We can put up six tomorrow, that's about all. I've asked for replacements—pilots *and* aircraft—but they'll be two or three days."

Two or three days. . . !

"Yes, of course."

"We're putting in the rest of those new windscreens tonight."

"Grand, sir. Saved my bacon today, and no mistake."

The ghost of a grin from Bouchier, then he said: "You're short of a flight commander. Any ideas?"

"Yes, sir. Could you get me Brian Kingcome?" Kingcome, a flying officer, was one of the best pilots in

Tuck's old squadron, 65. He'd been trained at Cranwell, the school for permanent officers, and was a knowledgeable and immensely popular man.

"I'll see what can be done—it's a good idea. Well ... see you in the morning."

As the Group Captain left, Bartley came in and claimed his free beer. He flourished a report from the ground crew chief, signed and witnessed, which proved that his aircraft had twelve wounds more than Tuck's.

"Fair enough," Bob said, "I'll lay you two to one for tomorrow night."

7

Before dawn next day his aide, Thomson, wakened him with a mug of tea black as anthracite. The face that peered at him from his shaving mirror didn't look like a squadron leader's face. It looked frighteningly like his own.

The tumultuous events of the previous day had changed his whole status, his purpose and responsibilities, but it would be some time yet before he grasped the facts, absorbed the sudden wad of experience and adjusted himself accordingly. Meanwhile—today, in just an hour or two—he would have to lead 92 into battle. He was hideously conscious that he lacked the easy charm, maturity and authority of Roger Bushell.

He was not afraid of the fighting that lay ahead, nor of the long odds against his own survival: he was afraid only that his leadership would prove faulty, that the other pilots wouldn't be able to give him their complete trust.

Then, wryly, he reflected that he wasn't really a squadron leader—only a half-squadron leader, for he'd be lucky if there were six or seven Spits serviceable this morning. . . .

The batman* Thomson, with his white temples and suffering face, was even more morose than usual. He spoke hardly a word and made a great business of finding tasks which kept him down on his knees or bent double—so that he didn't have to look at Tuck. As he left for the mess Bob was tempted to say something

*Orderly.

which would cheer up the old fellow, but Thomson was such a conscientious and sensitive character that he might easily take either a crack or a lighthearted reproach the wrong way. So Bob just said: "Thanks, Thomson. See you later."

The atmosphere at breakfast wasn't at all tense. One or two of the pilots seemed a little quieter and more thoughtful than usual, but generally speaking there was, if anything, a slight air of overcasualness. Nobody seemed to notice the empty chairs. There was no mention of yesterday's losses, though everyone had read a signal received from the Air Officer Commanding 11 Group: "Congratulations to 92 Squadron on their magnificent fighting and success on the first day of war operations. . . . The A.O.C. sincerely hopes that the squadron commander and other missing pilots will turn up later as so many others have in the past two weeks."

As they ate they heard their aircraft engines coming to life in the bays a few hundred yards away—being run up by the ground crews, who had worked all through the night to repair the ravages of the first day's fighting. Tuck, always a lusty eater, managed to scrounge a second helping of bacon and eggs.

As he was leaving for the flight, Bouchier phoned to say he'd asked for Kingcome to be transferred from 65, and he thought it would be all right. But there didn't seem much chance that the move would be approved until late that day.

"That's all right, sir," Tuck said. "I've got Tony Bartley—he put up a helluva good show yesterday. He'll cope."

"Yes, let him have a go, then. Incidentally, we'll manage to put up eight kites."

"Wizard—that's more than I expected."

"Better get over there pretty smartly. You'll be off any time now."

"Right, we're on our way."

They were airborne at five minutes past eight, bound for Dunkirk again. The weather was glorious—the sky was a gold glare that made the eyelashes stick together. The Spitfires' shadows skimmed over the rolling downs and then the blue rink of the becalmed Channel. The bridge of boats had thickened now—craft of all sizes and occupations crazily intermixed in the endless to and fro procession. One group of trawlers towed barrage balloons which marched sedately, like a herd of rhinos, five hundred feet up in the clear air. Others, returning heavy-laden from the battered strip of coast, trailed dark oil stains, smoke, broken rigging.

Tuck took the squadron up to 15,000. He had a new aircraft—3249. It felt young and eager in his hands, and soon, with startling suddenness, he found himself cursing because here they were over the beaches and there was no sign of an enemy formation.

This flaring aggressiveness surprised him: it seemed completely natural—it had come without any encouragement or screwing-up process, like the calm and deliberate zeal of a hunter once he has shouldered his gun not to come home empty-handed; and it poured confidence into him so that he lost his qualms about this new responsibility which had developed upon him so unexpectedly. He settled down to fly and to fight, and all else was forgotten.

As he looked down at the flames and the chaos and the struggling men, the fine bones of his jaws and chin suddenly showed sharp and hurtful under the tanned skin, and his voice over the r/t assumed that clipped, almost Asiatic flatness which was to become so well-known on the furiously crackling ether in the months to come. His lean body remained relaxed, moving smoothly, pliantly, but inside . . . inside it was as if the core of him went rigid, flint hard, sternly holding him to his task—and to the vow he'd made years before, after the air collision which had trapped him in the Gladiator cockpit: *never to let fear approach again, to hold it away by sheer aggressiveness and superconcentration on the job of flying.*

"Open it up—good and wide! Everyone keep his eyes peeled."

Slowly, uncertainly, the squadron fanned out until the Spits were sixty or seventy yards apart. It was against all the edicts of their training, but Tuck could see no point in having his boys preoccupied at a time like this with the mere business of keeping a neat, tight formation. Eight lookouts were better than one.

"Now then—keep your lips buttoned unless you've something bloody important to say. And when you *have* to talk, give heights and directions properly, and for Christ's sake don't just screech like a bunch of schoolgirls."

In silence the eight pilots flew on down the coast, each one searching the sky above, below and on all sides. Now every man felt keyed up, tough, efficient. Tuck had shown them he had definite ideas of his own, and that was impressive, comforting—for that was what made a good fighter leader.

They went up and down their "beat" twice, and then they spotted a formation of twenty Dornier 17s

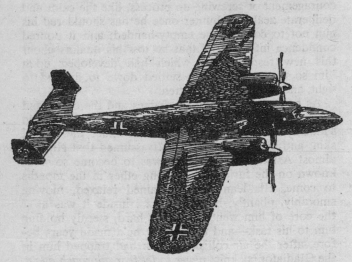

at about 12,000 feet, far inland, heading towards the beaches to bomb the cornered troops and the evacuation fleet. Behind the bombers, and about 7,000 feet higher—4,000 feet above the Spits—gleamed a protective arrowhead of Messerschmitt 110s. Tuck knew he had to ignore the fighters and try to break up the Dornier formation before they could start their bombing run.

"Buster!"

The eight pilots pushed their throttles forward through the "emergency" seals to full power and the Spitfires slashed inland in a shallow, curving dive. Tuck meant to bring them round in a wide half circle on to the tails of the bomber stream, but was prepared to change direction and risk a head-on or beam attack if the escorting 110s looked like dropping down to intercept.

A wonderful stroke of luck relieved him of this worry. As he glanced again at the high wedge of fighters he saw a squadron of Hurricanes hurtling almost vertically down out of the remote blue, quick as minnows, in a perfect "bounce." In mere seconds the Messerschmitt pack was shattered into writhing fragments, completely taken up with the business of its own survival. Now the Dorniers would have to look out for themselves.

The bombers were flying in wide, flat "vics" of three. As the Spits came down on their tails they were in a gentle turn to starboard, lining up to start their bombing run. Bartley's section, on the inside of the Spitfires' curving dive, got within range before Tuck's flight. As Bob recalls it, Tony did "a rather extraordinary thing."

"He went down the starboard side of the stream, shooting them up one wing, and I distinctly saw him leapfrog over one 'vic,' under the next, then up over the third—and so on. He did the whole side of the formation like that, and he tumbled at least one—maybe two—as flamers at that single pass. It was just about the cheekiest bit of flying I'd seen. The chaps in his section tried to follow him, but they managed only one or two of the 'jumps.' Tony made every one.

"Tony was a real goer. In those shaky first days he was a tremendous example—he went at the job beak and claws, and was always bubbling with gaiety. I don't think his value has ever been fully recognized."

Tuck, leading his section down to rake the port side of the stream, impulsively decided on different, equally unorthodox tactics. He closed his throttle and let down his flaps, using them as air brakes to slow down his machine so that he would have more time to take aim and could fire longer bursts. Leveling out from his long dive at high speed, there was every chance that the flaps would be ripped off, but he thought the risk was worth taking: the Dorniers had to be stopped before they reached their target, so the more bullets they could pump into them at this first pass, the better.

As the "flapjacks" forced the flaps out from the trailing edges of the wings into the raging airflow the Spitfire shuddered and pitched, as though she had become enmeshed in an invisible net, and he was jerked forward hard against his harness. But the flaps held, and his speed fell off rapidly to almost 180. He selected flaps "up," gently adding throttle and swiftly trimming the controls. Now he was able to hold the speed constant and approach the hindmost Dornier of his line steadily, comfortably.

He opened fire at about 400 yards. His bullets tore into the hindmost Dornier's port engine, wing-root and fuselage. He had managed to slow down enough to stay right behind it and keep on hitting it until it pulled away to the left, out of the formation, wallowing about and shedding lumps of wreckage.

The upper-gunner blazed defiance, but despite the Spit's low speed got no hits. Gunners from other Dorniers on Tuck's right quickly laid on a heavy crossfire, and then bullets came snickering into the cockpit around his feet. A searing pain in the inside of his right thigh made him jolt in his seat, but the leg still moved easily, so he stuck to his task.

The stricken bomber's port engine began to vomit black liquid. It banked and turned sharply away from its companions. Tuck followed round and dropped below it, gradually closing the range until he was firing up from about 100 yards into its belly. The underside was painted pale, duck-egg blue, and it looked soft, vulnerable. At this low speed, and with the recoil of his eight Brownings, the Spit's controls were getting sloppy and he had to work hard to fly her with any precision. Yet he managed to put two long bursts into the belly before he was obliged to add power and break off to avoid stalling.

Even as he drew away to one side, he saw two of the crew bale out—hunched, flailing shapes flashing down ahead of him . . . white streams sprouting as he passed over them, and then the filled parachutes diminishing rapidly far behind. He waited for others —the Do. 17s were believed usually to carry four— but none appeared. Suddenly the whole port side of the bomber burst into flames. It went over steeply, almost on to its back, and started down vertically, cutting a long, black scar in the sky.

At once he gunned his engine and headed back for the enemy formation. The Germans were fighting gamely, trying to hold together, but they'd lost several machines and their "vics" were becoming very ragged. The Spits were tearing into them again and again and it seemed that with any luck the stream would disintegrate very soon, abandoning its mission. Already the leading Dorniers were swinging their noses gradually round towards the north—away from their target, getting ready to dash for their bases.

He picked out one, again on the port side, and clobbered it from about 400 yards dead astern. At once the bomber stuck its nose down and, holding to the same northerly course, dived steadily at full throttle. Bob followed directly behind it, firing short bursts, glancing every few seconds into his rear vision mirror to ensure that he wasn't bounced by a stray 110. Not a single bullet came back at him. There was no trail of oil or coolant, no wisp of smoke. No one bailed out. Its speed mounted, but its direction and angle of dive remained unaltered until it hit some sand dunes and exploded in a fountain of orange flame.

Almost out of ammunition and very low on fuel, Tuck turned for home, crossing the Channel at about 500 feet. A throbbing in his right leg reminded him he'd been hit. He found his thigh was sticky with blood and did his best to staunch it with a handkerchief. He didn't feel giddy or sick, so he told himself it couldn't be serious.

At Hornchurch he found only a few marks on 3249 —some bullet holes in the bottom of the cockpit and a groove along the top of the canopy. "Free beer for the workers!" yelled Bartley, proudly displaying a riddled tailplane as proof that he'd won the bet again, and encompassing with a sweep of his arms the team of sweating fitters slaving to repair it.

Tuck's leg was very stiff now, and when he got out of the cockpit he couldn't quite straighten it. He noticed a small tear near his right trouser pocket, felt inside and fished out from his loose change a pocked and buckled penny: this coin had stopped one bullet, but another must have lodged in the back of the thigh. Carefully he placed the damaged penny in his breast pocket, then hobbled to the flight hut leaning on an airman's shoulder and let the M.O. take a look at the wound while he reviewed the action with Bartley and Allan Wright.

Peter Cazenove was missing. Somebody had seen him going down close to the French shore.

(It must have been at just about this time that Cazenove, having landed safely on the dunes near Dunkirk, walked along the beach and came upon the gutted shell of Pat Learmond's aircraft. Rummaging in the wreckage, all he could find was the blackened, twisted buckle from Learmond's parachute harness. Cazenove joined up with a party of the Rifle Brigade, fought with them in the last hours of their glorious rearguard action at Calais and was taken prisoner.

The M.O. found a small, deep hole on the inside of Tuck's thigh.

"Something in there, all right—in the muscle, I think —and it'll have to come out bloody quick. C'mon— into my car."

They drove to Romford General Hospital—because the doc thought this might prove tricky, and they'd need X rays—the hospital had more equipment than the station sick quarters. Tuck was still wearing flying boots and a gaudy silk scarf, and it was obvious that he'd come straight from his airplane, but a vinegary nursing sister brushed aside their requests to see the casualty surgeon immediately.

"In here!" she commanded, opening white double doors. "You'll have to wait your turn like everyone else."

They passed into a large waiting hall with row upon row of benches, packed with ailing humanity. There were small boys with stys, expectant mothers, factory workers with bashed fingers, old men with boils, dozens of people of all ages wheezy and flushed with colds or drooping with aches or fevers. Few of them gave Tuck a second glance. All sat in glum silence, dulleyed, preoccupied with their private sufferings.

The M.O. waited just thirty seconds, then walked over to one of the doors at the far side of the hall, rapped smartly once, strode in. Scores of heads turned to watch the door for a suspenseful minute or so, until the M.O. thrust his head out and beckoned to Tuck to join him. Then, as Bob hobbled past the crowded benches there came a resentful stirring and muttering: he felt his lips tighten and his cheeks go hot, and he didn't dare look to either side. The M.O. held the door wide, and when Tuck had entered he stuck his head out again, sniffed loudly, then closed it with an eloquent slam.

The hospital surgeon examined the wound and said he'd have to probe.

"I don't want it stiff like this, that's all," Tuck told him. "Do anything you like, but don't leave it so that I can't bend or straighten it."

The surgeon nodded, understanding only too well how important it was that this particular leg should be sound and usable again within a very few hours. He was a middle-aged man with a tired, drawn face and the deft, pink hands of a bartender.

"I don't think it's touched the bone," he said. "The stiffness is just bruising. Quicker it's out, the quicker it'll wear off."

The probing was painful, but brief. In less than two minutes the surgeon held in his forceps a tiny square of metal. It wasn't a bullet, nor a piece of shrapnel: it was a small duralumin nut from the rudder pedal! It must have been knocked off by one of the bullets which had raked the bottom of the cockpit, and it must have been traveling with terrific force to embed itself so deeply in his flesh. Tuck took it, and put it in his pocket. Along with the bent penny.

"A few inches higher," the surgeon remarked as he sterilized and dressed the wound, "and they'd have had to transfer you to the W.A.A.F."*

Driving back to the airfield, he kept flexing his knee and sure enough the stiffness began to go. As they swung in through the main gate the M.O. said quietly: "Those people in the waiting room—don't blame them too much. The public hasn't really grasped what's happening across the Channel yet. Most of the details are being censored—till the evacuation's complete. To them the war's still far away."

*　　*　　*

Bouchier was waiting for him.

"How's the leg?"

Tuck told him, and showed him the nut.

"Glad you're fit. I'm sending you up to have a talk at Air Ministry."

"Now?"

"Yes. It's very unlikely there'll be any more flying to-day. Unless there's a first class panic I'm standing down 92 until the spares and replacements come through. Get off soon as you can, will you?"

"Yes, of course. Whom do I report to, sir?"

"They'll have a car waiting at Northolt. The driver will know where to take you."

On the tarmac one of the undamaged kites was waiting, engine running. He swigged a mug of coffee and five minutes later was in the air. The underside of his thigh was tender and painful under his weight, but the stiffness had almost gone now.

He was relieved and happy. In his first twenty-four hours of fighting he'd destroyed five enemy aircraft, and managed to bring home a severely damaged Spitfire —so that it could be patched up and be put back into service in a few days. A nick in the leg wasn't worth considering as a debit item.

From Northolt the Ministry car whipped him to Whitehall. He was escorted to the office of Air Commodore Donald Fasken Stevenson, Director of Home Operations. Stevenson and a collection of staff officers

*Woman's Auxiliary Air Force.

were huddled over maps and sheaves of reports from the fighter stations. They all looked as if they hadn't slept for about a week.

Stevenson regarded him with eyes glazed and bright as painted porcelain, nodded him into a chair. Speaking softly, and with great charm, he came straight to the point.

"We need information—first hand stuff, from a chap who's been in the thick of it and knows what he's talking about. All of us here are going to fire all kinds of questions. Please keep your answers short, and if you can't help on some points don't waste time on them —just say you don't know and we'll get on to something else.

"Let's start with generalities. From your angle—that is to say, as an operational pilot without access to overall statistics—how do you think it's going?"

"Pretty well, sir. Their losses are much heavier than ours—I'm quite certain of that. But of course, they have terrific superiority in numbers."

"What about the chaps—do the numbers worry them?"

"Sometimes they get a bit niggled—er—angry, that's all."

"Not depressed—jumpy?"

"Not at Hornchurch, sir. Everyone there's pretty pleased with results so far."

A wing commander of the Intelligence branch broke in.

"Your squadron's had heavy losses—including the C.O.?"

"That's so."

"All in a matter of a single day, eh?" Tuck nodded.

"Didn't that shake the younger pilots?"

"Probably. I know it shook me."

The wing commander looked momentarily flustered, and Tuck was instantly sorry for him—he hadn't intended to embarrass a ground staff officer like that; his answer had been completely sincere.

"Would you go so far as to say your chaps are still keen?" another wing commander asked.

"Keen? Yes, I would. In fact, I'd say they're keener

and steadier than ever now, after mixing it once or twice. I don't think you need worry about the chaps, gentlemen—what *would* help would be more replacement aircraft." At that several of them sighed, and some of their shoulders drooped a little.

"The next biggest difficulty is refueling," Tuck got in quickly. "That's going to get more and more serious if Jerry keeps forcing the pace. Obviously, once the evacuation's complete he'll be over here, trying to shoot up our bases.

"It seems to me vital that we must be able to service our machines and get them back into the air just as quickly as possible—and that means a lot more gas tanks. Every minute that an aircraft spends standing on the tarmac increases its chances of being written off by a strafing intruder.

"The fuel trucks we have now are hopelessly out of date. They only have putt-putt engines and sometimes there are only two of them to a full squadron. That means waiting in line for juice—and the turn around may take ninety minutes or more."

They listened patiently, but their faces showed that they'd heard all this before, and that there was no immediate hope of improving the situation. All the same, he meant to press home his point, so he took a deep breath and said in a slightly quieter tone: "We won't really be getting the best out of the pilots and machines we have until we get a truck that can give us rapid refueling. And then I think there should be one truck to each machine."

More significant sighs, and some heads shaking. Stevenson, who had made a few notes while Tuck was speaking, pushed a cigarette box across his desk.

"All that's good sense, but the problem isn't going to be solved in days or weeks. Meanwhile we must decide how best to carry on with what we have at our disposal. So I think we'd better move on to something else now."

The questioning lasted an hour and a half. They moved on to discuss the efficiency of other equipment, methods of communication and ground control, and kindred technicalities. Tuck gave them his frank opinion that ground control was still very poor, but didn't try to push any more of his own ideas for improvements.

Nor did he mention his pet theories about looser formations and—an idea that had come to him during only the last few hours—the need for squadrons to join up and operate as wings of up to thirty-six aircraft. It seemed to him that if squadrons could operate in close partnership, it should be possible for some of them to draw off the German fighter escorts, leaving the bombers for the others to tackle. But he realized Stevenson and his staff were concerned only with the broad, pressing problems and had no time to ponder purely tactical matters which could best be handled by station commanders in the light of battle experience.

*   *   *

Kingcome joined 92 the next day, but though the squadron stood by from before daylight, apart from a twenty-minute wildgoose chase by Tuck and Titch Havercroft over the Thames Estuary in search of an "unidentified aircraft"—which turned out to be an amiable old Anson trainer on a navigation exercise—they spent almost the whole morning hanging around at dispersal. The weather was clear and they knew other squadrons were up, mixing it over the Dunkirk sector.

Tuck alone guessed the reason for their inactivity: the spares and replacements hadn't come through yet, and Bouchier was loath to send 92 into battle in its weakened state unless things got desperate. Things got desperate about 11:30, and they were "scrambled." As far as Tuck remembers, they were now only seven aircraft. As they climbed, control gave them the expected order: "Offensive patrol, Dunkirk-Calais-Boulogne."

They patrolled for the best part of an hour at about 20,000 without sighting an enemy machine, then Kingcome spotted a single Dornier 17 a few miles inland. It was almost certainly on reconnaissance, coming out to photograph the beaches. It wheeled and darted for home as the seven fighters pounced.

A reconnaissance Dornier 17 with its nose down could do almost 300. This one trailed light blue smoke as its pilot rammed his throttles and mixture levers forward, overboosting madly in his bid to escape. In the race to overtake it Tuck and Kingcome led all the way, almost abreast of each other. As they came into range the German rear-gunner began a stubborn defense, swinging his sights smartly from one attacker to the other, firing short and fairly accurate bursts.

The bomber took no evasive action. Its pilot merely kept his nose down and his two engines screaming full bat. Tuck scored the first hits, and knocked out the gunner. Brian had a go, then Bob closed in and started to batter it hard. Pieces flew off, and suddenly the three remaining members of its crew bailed out.

Tuck kept on pumping bullets into it, but it flew on in its dead straight dive, engines still roaring at top revs, without a flicker of flame or wisp of smoke. He could see his fire actually passing through it, but it refused to blow up. He drew up alongside and had a good look: between the tail unit and the cabin there seemed more daylight than fuselage—the Dornier was a flying sieve. On the r/t he could hear Kingcome's spluttering laughter and, glancing in his rear vision mirror, he saw the entire squadron strung out in an orderly echelon behind him.

"G'on Skip!" somebody yelled. "Throw your boots at it!" He had a strong suspicion that was Havercroft. . . .

He slid back astern of it and let fly again, from about 150 yards. More lumps flew off, but still it kept on, straight and resolute, as though running on invisible rails. They were down to about 4,000 feet, so he gave up and called the boys back into battle formation. As they turned for home they saw a bright flash on the ground well to the north: the "ghost" plane had finally blown up—in its own good time.

At seven o'clock that night the squadron was transferred to Duxford "for rest cure and maintenance." For the next few days, while the last of the evacuation craft limped into the southern ports and the Wehrmacht moved in to snuff the last pockets of resistance on the French coast, 92 was out of the fight.

On the 27th May Fighter Command, preparing for

the inevitable Luftwaffe attacks on its bases, began a wholesale redeployment of squadrons throughout the south of England. In quick succession Tuck moved his squadron from Duxford to Martlesham Heath, to Feltwell, then back to Duxford again. In between these moves they flew one or two patrols, but with the end of the Dunkirk phase in sight, a lull had descended and they saw nothing.

Tuck and Kingcome were becoming good friends. Brian, with his scarred, box-jawed face—the result of a car crash years before—had the sort of toughness and defeatless gaiety that Bob admired. Brian was "a man's man," fond of sports and beer, and he didn't have much time for women. They became constant flying and drinking companions.

News arrived one evening that Desmond Cooke, C.O. of 65, had gone down a "flamer." Tuck was deeply grieved, but he'd been expecting it. Only a few weeks ago he'd dropped into the 65 mess to have a few beers with his old chums, and Cooke's appearance had shocked him.

Desmond seemed to have aged greatly; he was thinner, desperately tired, and his eyes had a glazed, dreamy look. They chatted for a while together, away from the others.

"I dunno, Tommy," Desmond had said, drawing his fingers down his cheeks as if to smooth the fatigue from his face, "things seem to be happening too fast for me . . . I can't keep up, that's all there is to it. You know, this is damned funny—why am I talking like this? Haven't said a bloody word to another soul, and here I am suddenly spilling the beans to you. Why d'you suppose that is, eh Tommy?"

Tuck remembered that during his last few weeks with 65 he'd noticed Cooke slowing up. In practice "scrambles," the whole squadron would be in their cockpits with the engines started up, all raring and roaring to go—and half a minute or more would pass while the C.O. fiddled and fussed with his straps and his cockpit drill, eventually got his prop turning and started them moving. In the air too, his reactions had become sluggish, his mind indecisive.

Cooke, Tuck well knew, didn't lack guts—otherwise he'd have got himself the rest cure he obviously needed, and wouldn't be still at his post now. But the high powers and speeds of modern aircraft, the more complicated cockpits and the elaborate, constantly changing battle drills and r/t procedure were wearing him down. He was a very experienced and conscientious pilot, with a fine record, and he was still under forty: but somehow his mind couldn't grasp things quickly any more—he was becoming confused, uncertain. He was, in short, no longer suitable for fighters.

Tuck had tried not to show that he felt sorry for him, and switched the conversation to other topics. Desmond had talked quite gaily after that, and even laughed a bit, yet his eyes remained dreamy, far away. . . . Tuck had left feeling most uneasy.

And now it had happened—dear old, conscientious Cookie was gone. Tuck blamed himself for not persuading him to apply for a rest, or even a transfer—he would have made an excellent bomber captain. But on reflection he came to the conclusion that nothing he could have said would have made any difference: remembering that glazed, dreamy look, he felt sure that Desmond had known he would shortly be killed. And, with the highest sort of courage, he'd reconciled himself to the fact. . . .

Tuck didn't want to brood on it, so he started a party that lasted until 2 a.m., when the adjutant forcibly closed the bar. By then he'd forgotten all about it, and he fell asleep as soon as he reached his room—without taking off his clothes.

At 5:15 a.m. on May 28th, they were ordered to Martlesham Heath, "to rendezvous for offensive patrol." This meant they were to join with at least one other squadron. Tuck was delighted—the Air Ministry was proving more receptive to ideas than he'd thought possible.

As they landed at Martlesham a sudden summer mist was rolling in. By the time they'd taxied to the tarmac it was so thick that they could barely see their prop blades.

"It won't last," a ground crew corporal assured him. " 'appens 'ere quite often of a mornin', sir, but it always clears once the sun's well up. There'll be some tea over at Flight . . ."

But Tuck wasn't interested in tea. He walked up and down beside his Spitfire, peering up through the curling grayness at the distant splotch of the sun, hating every minute of the waiting. It was very still and chilly. He hunched his shoulders and stamped his feet.

Out of the mist a short, burly figure materialized, advancing down the tarmac with an odd, lurching gait that somehow seemed cocky, aggressive. The newcomer was a flight lieutenant. As he drew near Tuck saw he had piercing gray-blue eyes. A furry black pipe was clenched in his powerful, square jaws.

"What's the score, old boy?" the stranger demanded.

The words were fired out at great speed. The tone seemed positively belligerent. Tuck's own impatience flared. What had this "prune" got to act so tough about? Neither of the other squadrons waiting with 92 here had seen action yet. Tuck felt a kind of duty to Bushell, Learmond and the others.

"Haven't a clue," he snapped. "We're *all* waiting to find out, aren't we? It's all right—we know you're bloody keen types."

The two of them stood staring at each other, alone in the mist's dampness. After several seconds the stranger took the pipe out of his mouth, swore very profanely, and began to lurch away. As he swung round Tuck noticed that knotted round his thick neck he had what looked very much like a silk stocking, with one end hanging loosely outside his uniform. He couldn't resist calling after him: "By the way . . . if I were you I wouldn't fly with that thing round your neck."

The burly stranger stopped and looked back over his shoulder. They were three or four yards apart now, but those insolent eyes seemed to shine through the grayness.

"Why the hell not?"

"Because if you have to bail out that loose end's

liable to get caught up on something and you'll hang your silly self. If you must wear it, stuff it inside your tunic."

Tuck turned and stalked back towards his aircraft. Behind him he heard a low, indistinct snorting and muttering.

It wasn't until much later that day, in the Martlesham Mess, that he learned that the burly flight lieutenant was Douglas Bader, a flight commander of 222 Squadron. And he was shaken when it was explained to him that Bader's odd, lurching gait was due to the fact that he'd lost both legs in a crash years before, and now walked—and, by all accounts, flew very well—with artificial limbs.

"I remember," Tuck says, "that my immediate reaction to this rather obstreperous man who came blustering up to me that morning was that he was much too cocky and ought to be taken down a few pegs. Of course I didn't know then who he was. Extraordinarily enough, though Douglas was by that time something of a legend in the air force, I'd never heard about him.

"Yet, you know, even if I had known his story, I'm sure we'd still have sparked right from that first meeting. In fact we still spark, to this day. Now, of course, we do it in an affectionate way, because now we know each other. From the moment I got to know Douglas I've had the greatest respect for him both as a pilot and as a man.

"Whenever we run into each other nowadays, even if it's at some important function attended by distinguished guests, Douglas invariably bawls across the room 'Bobbie, you bloody old scrounger—go home before you get drunk!' Then I'll shout back at him, something like 'Belt up, you rude sod!' That's the usual form—it just depends which of us spots the other first."

On that day when they first set eyes on each other, Tuck and Bader flew in the same formation for the first—and only—time. Unfortunately, their recollections of the exact circumstances differ markedly.

Bader's biographer, Paul Brickhill in *Reach For The*

Sky, opens his account of the day with this brief description of the meeting between Douglas and Bob:

"Bader strolled over and asked 'the form' from a slim, handsome flight lieutenant, elegant in white overalls and with a silver name bracelet round his wrist.

" 'Haven't got a clue,' said the debonair young man, who had aquiline features like a matador, a thin black mustache and a long, exciting scar down the side of his face, the type of young blade, Bader thought, who would make a young girl think of darkened corridors and turning door handles. His name was Bob Tuck."

From this we can see very clearly Bader's first impression of Tuck—a highly erroneous one, for despite his unquestionable handsomeness Bob was much more interested in beer with the boys than girls: though by no means a monk, if there had been any turning door handles in his life, the door had been his, and the uninvited caller some "empty-headed, nuisance-making female." At any rate, we can understand how Bader's snap assessment, coupled with his natural aggressiveness, influenced his whole attitude on the tarmac that morning—and we can understand why Tuck's temper flared. Bob was the C.O. of a squadron which already had taken a terrible pasting: this gruffness and faint, though unmistakable, derision from an uninitiated stranger of subordinate rank naturally infuriated him.

Brickhill states that 222 were led off by Squadron Leader "Tubby" Mermagen "in four neat vics of three." Tuck's logbook shows that 222 operated that day as part of a wing, that the leader of the wing was Bob Tuck, and that they were spread out, flying in loose pairs.

On one point, however, they are agreed: the whole show was a wash out, and they paraded up and down the French coast for a hundred and twenty humiliating minutes, encountering not a single German.

On June 2nd, 92 were ordered to Martlesham again, but this time Bader's squadron wasn't in partnership —there were two others, newly brought down from the north of England.

Soon after 7 a.m. the wing of Spitfires flew to the Calais area and spotted eight Heinkel IIIs with fighter "umbrella" coming down the coast from Holland, with the sun behind them. The escorting wedge of about 25 Messerschmitt 109s was higher than the Heinkels and considerably behind them. Tuck, in fact, thought the fighters were much too wide of their charges, and there was a good chance of getting among the bombers before the escort could dive in to protect them. Accordingly he made a direct beam attack on the Heinkels, breaking up their formation at the first pass.

At this time Tuck had established an arrangement with Bob Holland. They flew as a pair, Holland protecting his leader's tail. They had worked out their scheme by talking into the wee small hours, and so close was their understanding that Tuck didn't need to give a single order over the r/t—Holland could tell exactly what maneuver his C.O. was beginning by the first tilt of Tuck's wings or dip of his nose, and he didn't fail to follow him faithfully now, through this swirling, leaping mêlée.

Tuck quickly fastened on to a Heinkel, which had narrowly missed colliding with one of its companions, and was now veering away from the others. One long burst shattered the upper-gunner's kidney shaped canopy. The second set the starboard engine ablaze. One of the occupants managed to bail out before it whipped on to its back and plunged vertically.

As he climbed he counted three other bombers going down in flames. And then suddenly Holland yelled: "Watch it—109s above you!" The German fighters finally had caught up with the running battle and six of them were dropping on Tuck. In the process one of them crossed in front of Holland, who promptly shot its tail off. Another took fright in mid dive, broke off and passed close to starboard of Tuck, going into a steep climbing turn.

He yanked the Spit round in a full bank and, trusting to Holland to let him know if another looked like getting on his tail, took his time in lining up his sights. A burst from about 300 yards raised a heavy glycol trail, and a few seconds later the 109 began to burn so furiously that he imagined he could hear it crackling.

A wing wilted away and it hurtled down in a vicious spin.

Only then did he realize that he'd received damage in his earlier attack on the Heinkel. The upper-gunner had put some holes through the Spit's prop and the nose. Fortunately the bullets had missed the radiator and the glycol tank, but the engine revs were vacillating uncontrollably. He dropped out of the scrap and circled until the surviving Germans had streaked out of sight, and then most of the other Spits swooped down and reformed on him. All landed safely at Martlesham.

The first "big formation" operation of the war, led by 23-year-old Bob Tuck, had been an outstanding success. But nearly all of 92's machines had suffered damage: once again they were ordered back to Duxford for maintenance. "If you don't drive more carefully," one of the Martlesham ground crew sergeants warned, "sure as hell they'll suspend your license!"

8

The Air Council reckoned they'd taken the first round. In nine days over the Dunkirk area—from May 26 to June 3—at least 377 enemy aircraft had been destroyed for a loss of 87 R.A.F. machines. Exactly four-and-one-third German machines for every plane expended—or, since the majority of enemy machines were multi-engined types carrying crews, about nine dead or captured German airmen for every British fighter pilot who did not return.

Long before Dunkirk the Luftwaffe's policy in fighting the Hurricane squadrons based in France had been to overwork the British pilots, to keep them in the air until they were exhausted and could fight no more. (One Hurricane pilot, Geoffrey Allard, was actually lifted from his cockpit at the end of a day's fighting, fast asleep.) This policy would probably be doubly effective now that the British fighter groups were concentrating in their own little island, their numbers already depleted, their men wearied by the unequal battles across the Channel.

The fact that the invasion barges, the bomber hordes and the fighter packs didn't come in the first days of June is still hard to explain. Most probably it was Hitler's blind faith in the occult, coupled with the Nordic love of extravagantly staged rituals, which led Nazidom into the incredible blunder of delay. For while they were busy practicing their triumphant fanfares, rehearsing their processions and fly-pasts, giving each other Iron Crosses and moving the French railway carriage in which the 1918 surrender had been signed to the identical spot in the forest of Compiegne—so that this time the roles would be reversed—the Royal Air Force was able to draw breath, repair the ravages of Dunkirk, settle into new bases and reorganize for the second, perhaps decisive round.

In this period of frenzied preparation, Bob Tuck was chosen to work on a project of the highest importance. In the workshops of the Royal Aircraft Establishment at Farnborough, scientists and specialists of the Air Fighting Development Unit had stripped down and then reassembled the first Me. 109E to fall into our hands. It had been captured intact after force landing in western France a few weeks earlier. Tuck and Wing Commander George Stainforth, a brilliant pilot who had set up many air records in the 'thirties, were ordered to fly an exhaustive comparison test, matching the 109 against a Spitfire. To eliminate any difference in skill between the two officers, they were to change machines half-way through the trials and repeat their program exactly.

Veteran fighter pilot Group Captain Harry Broadhurst and a large party of brasshats, backroom boys and Rolls-Royce experts were assembled to watch the tests. The "boffins"* cuddled their calculations, clicked their slide rules and chattered excitedly.

Stainforth took the Messerschmitt first. Tuck was in his own service Spit. They got down to their first task by forming in line abreast at about 20,000 feet, wing tip to wing tip, flying absolutely straight and level with their throttles fixed at pre-arranged settings. Then gently they eased their sticks forward into a shallow dive, without touching their throttles, to see which fighter would draw away. This called for extremely accurate flying, for if either pilot had the slightest skid or sideslip on, he would lose some of the effect of his streamlining—the machine would present a greater resistance to the air and the speed wouldn't build as it should.

Nothing in it—German and British plane dived neck and neck. They repeated the maneuver with dif-

*Technician or scientist.

ferent throttle settings and obtained more or less the same result.

In a flat out, straight and level race the German proved very slightly faster. In various rolls and turns, the Spitfire was decidedly more maneuverable.

When it came to pulling up out of a steep dive, the 109E had a most definite advantage. It could pull up much more sharply, and climb away a little faster. Then they turned to a problem which Tuck and most other combat-experienced pilots considered grave and pressing.

"At this stage"—Tuck talking—"109s were getting away from us fairly frequently by sticking their noses right down and going into near-vertical dives. This meant that the pilot had to take what's known as negative 'G'—in other words, in this maneuver he was on the *outside* of the curve, like someone going over the top of a switchback at terrific speed, so gravity's effect was to yank him up sharply out of his seat and throw him against his harness. In most other dogfight maneuvers, you understand, the pilot was on the *inside* of the curve which his machine was describing, and so he was subject to positive 'G', which rammed him down into his seat, bent his backbone like a bow, and pressed his chin forward and downwards on to his chest. Positive 'G' also could make him 'black out,' because as he pulled violently out of a straight path through the air his blood tended to rush to his feet. For a few seconds, in the tightest section of a pull out from a steep dive, his brain was drained of blood and consequently his eyesight blurred and momentarily he became unconscious.

"Well, we reckoned we were just as tough and fit as the Luftwaffe boys, and that we could take just as much 'G'—positive *or* negative—as they could. But to our dismay we'd found that our Merlin engines couldn't stand up to negative 'G,' whereas the Messerschmitt's Daimler-Benz seemed quite unaffected.

"This is the sort of thing that would happen: you'd work up behind a 109 at height, and just as you got set to blow him out of the sky he'd spot you, slam his stick fully forward and drop like a gannet, more or less vertically. If you tried to do the same, the moment your nose went down your engine would go 'pop-br-ang!', there'd be a puff of dark blue smoke—and you'd lost all your power for several vital seconds.

"We'd guessed the reason, of course, and the boffins who'd gone over that captured 109 confirmed it. The Daimler-Benz engine had direct fuel injection, whereas the Merlin had a carburetor which couldn't cope with the negative 'G' imposed by this sudden transition from the horizontal to the vertical.

"I'm pretty sure the German Intelligence knew this failing, and that the Messerschmitt squadrons had been briefed to adopt this power dive technique to get away from our Spits. All we could do in this situation was roll over on to our backs and *then* dive after them, but that took several seconds and the 109 usually had time to get well out of range or nip into the cover of a cloud. It was pretty damned frustrating.

"George Stainforth and I demonstrated this business, and firmly established the defect for the benefit of the Rolls-Royce chaps who'd have to find the solution. Within a few days they'd devised a floatless carburetor for the Merlin that functioned perfectly under the most violent negative 'G,' and in a matter of a few weeks it was fitted to every operational Spitfire. It shook the Messerschmitt lads when we began going straight on down after them and blasting them as they dived. They changed their tactics pretty smartly."

Dogfighting high over Farnborough, Tuck suddenly found Stainforth's 109 squarely in his sights. Staring at the now familiar black crosses, for one instant sheer instinct took charge and moved his hand to press the firing button. He checked the action just in time—not that it would have been fatal, for the Spitfire's guns, though loaded to establish full fighting weight, weren't cocked.

For the second run through the test program Tuck got into the Messerschmitt.

"Right away I realized one reason why the 109 could pull out of a dive more sharply than the Spit. The rudder pedals were several inches higher than ours—in fact, the pilot sat with his legs very nearly horizontal.

This, of course, considerably reduced the effects of positive 'G' in a pull-out, because with the feet high there wasn't the same tendency for the blood to drain away from the upper body. As a matter of fact, some weeks before I'd had my own pedals extended upwards for about six inches, and I'd found that I didn't black out so easily.

"The Medicos were very interested in this point, and they supported the suggestion that we should have higher rudders. But some technical type claimed that with the raised pedals there was a danger that with full left rudder on, the toe of the right foot, as it came back, would foul the petrol cocks on that side of the cockpit. That was absolute nonsense, because you never used full rudder in the air—you just couldn't get it on against the terrific pressures on the control surfaces, and even if you *had* managed it, you'd probably have flick-rolled and over-stressed the whole aircraft! I had flaming arguments on this score, and in the end they let me fly with the extended pedals. Later the idea was adopted as a general principle on certain marks of the Spitfire."

He didn't like the 109 cockpit much. "It seemed even smaller than the Spit's, and the pilot's vision was decidedly poorer. The hood and windscreen were certainly far more robust, but they had a lot of thick, metal strutting—heavily studded, like girders—in front and on the sides, and these obviously obscured several sections of the sky. Oddly enough, this one didn't have a rear-vision mirror.

"The instrument panel was very confusing at first because it was festooned with scraps of paper bearing conversions—from kilometers to miles, meters to feet and so on. I had to sit and study it all for quite a while before taking off.

"I was interested to note that the gunsight was a reflector job, not much different from our own. . . ."

Tuck's opinion, after repeating the test program in the captured enemy machine, was that the 109 was "without doubt a most delightful little airplane—not as maneuverable as the Spit, mind you, nor as nice to handle near the ground. It had a tendency to a rather vicious stall, because, you see, it was even smaller than the Spit. But certainly it was slightly faster, and altogether it had a wonderful performance.

"An odd thing that sticks in my memory: it had a distinct, peculiar sort of smell, a certain sourness that I've noted often since then, in every German aircraft I've got into. Like an empty beer barrel, or stale vinegar, maybe. Probably it was the type of dope or spray they used—I don't know. But it was unmistakable, alien. And not at all pleasant, not like the smell of our own kites."

After the Farnborough tests, he felt he knew his enemy. From now on, whenever he duelled with a 109 he would be able to put himself in his adversary's cockpit, to see in his mind's eye how the foe's hands were moving on the controls. He knew the Messerschmitt's capabilities—where it could beat a Spitfire, and where it was outclassed. He was deeply grateful for this experience. It greatly strengthened his assurance and had an important effect on many of the tactical ideas which had blossomed during and since his Dunkirk fighting.

A few days later he led Bobbie Holland and Allan Wright on a reconnaissance patrol over the sector Abbeville (8)-Amiens-Doulon. Each Spitfire now had been fitted with armored plating—one large sheet of it behind the pilot's seat. As they sped low over the enemy occupied territory, they realized what a comfort this plating was: they still experienced that sickening, twitching sensation at the base of the spine whenever tracer from an odd A.A. battery came hosing up after them, but it wasn't nearly so acute now that each man had his back against a wall of tough British steel.

It was late in the evening and the light was starting to go when, near Doulon (9), Holland flicked over his microphone switch and yelped: "Woops—look out behind!" Streaks of light flak, like strings of luminous pearls, were reaching up through the gloaming. As they opened up the "vic" and began to weave away from the fire, Tuck was hit. About two feet of his starboard wing tip went spinning off astern, and the stick was almost snatched out of his grip as the aircraft bucked and writhed and turned, momentarily knocked out of its purposeful path, like a swimmer caught unaware by a powerful breaker. He regained control quickly, and the Spit flew on with her wounded wing held slightly low, as though hunched in pain. But the little formation had been effectively broken up—to get out of the sudden spray of shells, Bobbie had executed a half flick roll down on to the deck and gone scudding away just over the hedges, while Allan had peeled off to one side and disappeared into the gathering dusk.

Tuck dived and joined up with Bobbie, but they couldn't contact Allan. They set course for home, staying down at about thirty or forty feet. As they flew up a long, gradual slope, more flak came curving down at them, this time from a gun niched in a big, long barn on the ridge ahead. In what seemed little more than a heart's beat, Tuck had swung his plane round and with cool precision laid his sights on the emplacement. Holland saw his leader's burst tumble the gray-uniformed gunners from their posts. For two or three seconds the whole interior of the barn was filled with flashes and hazy smoke, and then as the two Spitfires passed over the roof both pilots glimpsed a flicker of flame. Probably the hay in there had caught alight. They decided they were justified in claiming the gun and crew as a "write off."

A few minutes later they came upon a column of military trucks heading west. They pulled up to about three hundred feet and, one on each side of the road, raked the entire length of the convoy. Now the dusk was thickening, but Tuck saw one of the trucks swerve into a ditch, spilling out soldiers. All along the line men were leaping from their vehicles, diving for cover. Some of them seemed to be rolling about on the roadway, like clowns tumbling on a circus mat.

Tuck wanted to circle round and make another pass at the convoy, but Bouchier's orders had been clear. So they flew back to Hornchurch, found Allan Wright waiting for them, and then reported to the station commander.

Bouchier was intensely interested in the results of

their patrol. It had been, in fact, something of an experiment, because attacks by low-flying Spitfires on enemy ground installations and supply lines were not yet a part of Air Ministry policy. The planners were preoccupied with the business of organizing the defense of Britain, and they considered—quite reasonably—that every fighter aircraft must be conserved for this task. Limited, low level reconnaissance over western France was permitted, however, and Bouchier had taken advantage of this loophole to send three of his best pilots over, hoping fervently that they would run into something worth shooting at. He'd thought it important to find out, right now, just how effective a Spitfire's guns were against A.A. emplacements and motor columns. He was delighted with Tuck's report, and passed every word of it on to Air Ministry, marked "Top Priority."

Tuck wanted to organize more low level sweeps, but couldn't get permission. Weeks dragged by, suspenseful and rumor-logged. Italy's declaration of war increased their impatience for action, but still no big enemy formations appeared over the Channel.

Then in the middle of June they were moved to a new base—Pembrey, in South Wales. They were furious, for it seemed like a regiment being taken out of the front line in disgrace. Despite all assurances, they believed they were no longer rated as a first class squadron.

Tuck's biggest sorrow was leaving "Daddy" Bouchier, for whom he had not only great respect, but much affection. They had known each other since 1937, when Bouchier, then a squadron leader, had commanded an outfit which shared Hornchurch with 65. Their first personal contact had come when Bouchier, acting as station commander while "Bunty" Frew was on leave, had to place Tuck under close arrest pending court martial on charges of low flying. Bob had "beaten up" the 'drome in a Gauntlet on a day when the rest of the planes were away on air-firing exercises, and he knew there were no senior officers about. Unfortunately his hair raising display terrified an airman's wife, eight months pregnant, who watched from a doorway of the married quarters. She collapsed and gave birth to her baby several days ahead of schedule. Mother and child were perfectly fit, but the hospital authorities had reported the facts to Bouchier.

Tuck made no attempt to deny the charge and he was duly tried and sentenced. The penalty was surprisingly light: a severe reprimand, and forfeiture of three months' seniority.

As usual, the court had been made up of officers from other stations, but "Chad" Giddings had served on it as "member under instruction." From something Chad told him afterwards, Tuck had good reason to believe that Bouchier had intervened and asked the President of the court to impose "the lightest possible sentence."

This affair had puzzled him deeply, for at the time Bouchier had seemed to him a fiery and unbending brasshat—certainly he had acted that way when Tuck had first been brought before him and placed under arrest! Besides, though temporarily acting as station commander, Bouchier had a squadron of his own— 54—to look after and had no reason to be much concerned about the reputation of 65.

Now, three years later, Tuck wasn't puzzled any more. He thought he knew exactly why Bouchier had intervened. All good squadron commanders, he had come to realize, wanted to see in their pilots two quite contradictory qualities: the spirit of recklessness, and a strong sense of discipline. Obviously these qualities couldn't always balance exactly to cancel one another out—occasionally the recklessness would burst through and dominate. Therefore, provided he was sure the culprit wasn't just an irresponsible fool or an intractable rebel, the wise squadron commander took care that the punishment didn't destroy the youngster's zest and dash.

Since his return to Hornchurch with 92, Tuck had come to know Bouchier as well as any man could. The station commander—"Daddy" now, with his iron gray, close-cropped hair and fatherly concern for his young pilots—still had a gruff voice and at times displayed a fiery temper, but he was one of the best loved "bosses" in Fighter Command. He drove himself unsparingly, and when his boys needed something badly, frequently would brave the wrath of his superiors by savagely slashing through red tape.

About this time the Germans began their first large-scale raids on Britain—by night. They attacked several airdromes in the south without inflicting serious damage, but killed and injured scores of civilians in towns and villages from Kent to the Scottish Lowlands. On the night of June 18th Sailor Malan, operating over Essex, bagged two bombers, opening fire in each case from only fifty yards range.

Tuck, flying uneventful patrols in the pastoral quiet of Wales, was thoroughly annoyed to hear of his friend's adventures. Ninety-two's situation seemed to make a mockery of Premier Churchill's speech, broadcast less than two weeks before—*We shall fight on the beaches; we shall fight on the landing-grounds; we shall fight in the fields and in the streets, we shall fight in the hills.* . . . All very fine and stirring, but here were experienced air-fighters stuck away in the western backwoods when surely they should have been manning the eastern and southern approaches.

The proprietor of the Stepney Hotel, the pub in the nearby township of Llanelly where they spent many of their free evenings, was a jovial Welshman inexplicably named William Maloney. (This name appealed greatly to the 92 pilots, most of whom had started flying before the war: in the pre-war air force a mythical band of pixies existed, known as "the Maloney Boys," who got the blame for engine failures and all minor accidents. These mythical creatures were the forerunners of the war-time "Gremlins," so widely publicized by the British press and radio comedians.) Maloney promised them a free bottle of champagne—an excellent vintage, too—every time they shot down a Hun. They never thought they'd hear a cork pop, and they made it clear to the well-meaning landlord that they didn't consider his joke funny enough to bear much repetition.

But within a few days they came to see that the move to Carmarthenshire wasn't so daft after all. Odd single raiders and recce. machines began probing up the Bristol Channel, and prowling around the busy ports of Cardiff and Swansea. Other, peculiar types of aircraft—big flying-boats, and long-range patrol bombers —suddenly became active along the coast of the neutral Irish Republic, and Intelligence believed these were trying to find and photograph quiet coves and estuaries where German agents could be landed. Despite war-time restrictions, it wouldn't be hard for spies to make their way from Ireland into Ulster, and hence across to England. . . .

Several times the 92 pilots, following the ground controller's directions to intercept an enemy prowler, found themselves a mile or two off Dublin, Wicklow or Tramore.

Once or twice the Irish guns fired a desultory, inaccurate volley at British aircraft. The only man who lost his temper at this was a young sergeant who hailed from County Cork. At the moment when the shells came up he could actually see his aunt's house near Rosslare. Two other pilots, Englishmen, were with him, and the flak puffs were a good mile away.

"Holy Mother!" his companions heard him cry at the landscape in general. "Now, is that the best yez can do? Are yez out to shame us, or what?"

One morning in the first week of July, Tuck and Bobbie Holland opened up from about 500 yards on a lone Dornier 17 on reconnaissance south of Bristol. They saw their fire striking home on the wings and fuselage, but it got away in cloud. In the mess that night Tuck declared: "If those had been 20-millimeter shells that clobbered it, instead of just .303 bullets, we'd have had a kill, and we'd be swigging Maloney's bubbly now! Why the hell don't they give us cannon?"

They gave him something else—a Distinguished Flying Cross, with a citation that praised his initiative in taking over command of the squadron after the loss of Bushell, and his personal example during the subsequent battles. (One or two of the phrases seemed to him oddly like those Bouchier used in reports and directives.) Another citation in the same issue of the *London Gazette* awarded the same decorations to Sailor Malan.

* * *

After dark one drizzly evening Tuck stalked a lone Ju. 88 for close on thirty minutes. The enemy pilot knew he was being hunted, and so far as possible remained cloistered in the cloud which was piled high over the Welsh hills. But at length he had to come out —to flit across a small area of clear sky from one patch of cover to another. He was in the white bite of the moonlight for only a few seconds, but Tuck pounced.

There was time to get in only one burst, from pretty well maximum range. He thought he'd hit it, for just before it disappeared into the overcast again, very distinctly he saw a cluster of black dots fall from its belly. At this distance his bullets couldn't have inflicted serious damage, but it was enough to make the Ju. jettison his bomb load and abandon his mission— probably a strike against Cardiff or Swansea. The Spitfire's fuel was getting low, so Tuck returned to Pembrey.

Very early next morning, he got a long distance telephone call. It was his father, speaking from home. His voice seemed lower, slower than usual.

"Robert, bad news I'm afraid. It's Peggy's hubby, John. We've had a telegram from his C.O. He was killed yesterday. D'you think you could get home for a day or two?"

"My God, yes . . . I'll ask for leave right away— probably fly down tonight. How's Peggy taken it?"

"Your sister's a brick, but it's been a dreadful shock. The doctor's given her something, and your mother's with her." Captain Tuck paused, cleared his throat, then said, "Well . . . if you can make it, Robert, I'm sure it would do her good to see you."

"Righto, I'll come tonight."

Half an hour later, patrolling at 15,000 over the Irish Sea, he was still trying to absorb the brutal fact.

Peggy Stanford Tuck and the gentle, jovial John King Spark had been married only thirteen months. John, 23, had been a Territorial, called up at the very start and posted to the Queen's Westminsters as a private; but recently he had been recommended for a commission and was expecting any day to be sent to an Officers' Training Unit. Somehow none of the family had thought of him as being in any sort of danger— at least, not for some months to come. Something very unexpected must have happened, Tuck concluded—he must have been whipped overseas on some hush-hush raid, or else involved in some stupid accident during training. . . .

And then he remembered something his mother had mentioned in a recent letter: John had been moved, suddenly, to an Army camp in South Wales, "a very lonely place, and rather primitive, where it always rains."

South Wales.

Terrible suspicion tightened its icy grip, a clenched fist in the pit of his stomach. No, it couldn't happen —the odds against it were too impossible! And yet . . .

When he got back he sprinted to the flight hut, grabbed the phone and called the Intelligence officer.

"Listen, old man, this is urgent. I want to know what enemy activity there was over Wales during the whole of yesterday."

"I can tell you right away, hang on." Rattling of a drawer and rustling of papers, then: "Here we are, sir. Very quiet day. Only one minor incident. A stick of four, late in the evening, near Porthcawl (10), Glamorganshire. Hey, wait a minute!—that must have been from the 88 you took a crack at—yes, time coincides!"

"What about . . ." Tuck's breath wouldn't come evenly, he had to force it out. "What about Porthcawl? What was hit?"

"Oh, nothing much. The stick fell in open country, but one was just on the boundary of the Army Camp at St. Donat's Castle. Damaged a lorry and a Nissen hut."

"Casualties?"

"Now look here, sir, you've nothing to worry about. If that Hun had got through to Swansea or Cardiff the chances are . . ."

"Were there any casualties?"

"Well, if you must know. . . . One private killed."

Tuck dropped the phone and blundered out of the hut, walking fast, blindly, his mind in ghastly upheaval. It was fantastic, but he recognized the authentic slap of Fate.

He turned off the perimeter track, and thrashed through some long grass, not knowing or caring where he was going. He fumbled a cigarette alight and tried to think calmly.

The odds against it were millions to one. If that bomber had taken off from its French base just two seconds earlier, or two seconds later . . . if the wind had been a shade stronger, or lighter . . . if the Ju. pilot had flown a hundred feet higher, or lower . . . if that break in the clouds had been smaller, or larger . . . if he, Tuck, had held his fire for one more second, or opened up a moment sooner . . .

If. He could go on on that theme for the rest of his life.

But the astounding fact had to be faced, had to be lived with: out of all the millions of people in Britain, those jettisoned bombs had killed John King Sparks. Therefore Tuck was responsible for the death of his own brother-in-law—it was Tuck who had caused the bombs to fall.

All just bad luck, you might say. "Tuck's Luck"—in reverse. But there was no getting away from the fiendish fact. Well, what was to be done?

Nothing. That was the answer. He couldn't tell his family—perhaps one day, but not now. It would only increase Peggy's agony. Women could never discipline their emotions enough to be logical. Not *that* logical, that was certain.

He finished the cigarette and started back. He had to arrange for leave and pack a bag. As he passed back along the perimeter track he never once turned his head to look at the lines of parked aircraft. For the moment airplanes had lost their fascination. It was the first time he'd ever felt this way. The first time he'd seen the downright ugliness of war.

But when he returned to duty after the funeral, three days later, he seemed quite his old self. He'd kept his dreadful secret from his sister and parents, and had even managed to push it out of his own thoughts.

From time to time he was ordered to attend conferences at the Air Fighting Development Unit, which now had moved from Farnborough to Northolt. At these meetings, whenever he got the chance he'd plug not only his old, pet theme about more and better gas trucks, but also his growing conviction that the Spitfire and Hurricane would be much more effective if armed with cannon instead of machine guns. He never seemed to excite much interest with these ideas, but he suspected that the same suggestions were being made constantly by other operational chaps, and the planners were already going into the pros and cons.

Flying back to Pembrey after one of these conferences, deep in thought about the problems they'd discussed, for once he failed to notice a menacing change in the weather. He entered a cloud and didn't realize how bumpy the air was becoming until the stick started kicking. Then he snapped out of his daydream and found himself caught in a violent summer storm. He tried to climb above the murk, but it stretched up, thick and turbulent, to well over 20,000. And its vastness was filled with jagged, blue-white flashes. His compass vacillated idiotically, and after thirty minutes more he realized he was utterly lost. For a pilot of *his* experience, this was very embarrassing. What a helluva thing it would be if, after all his striving for efficiency, all his success in battle and the winning of a D.F.C., he was beaten by a few minutes' forgetfulness and a whim of the elements!

He started calling on the r/t, using every channel fitted to the set, hoping that some aerodrome would answer. But all he could get in his earphones was a pandemonium of static, like a million whips cracking.

It was late in the evening, and soon the cloud that imprisoned him turned from dark gray to black as the last of the daylight went. For all he knew, he might be flying in circles over the Welsh mountains, or heading out into the Atlantic. This was the loneliest feeling in the world.

The storm pummeled him, its raging convection currents jerking the plane up and down like a yo-yo, the searing lightning and the sudden changes in altitude and air pressure making his stomach ache and his eyes blur. At all costs, he must keep above 3,000 feet—if he *was* over Wales then jagged peaks would be buried in the belly of the cloud. . . .

In the eerie, green glow of the instrument panel he sat and for over ninety minutes fought the wild night, with the cold sweat running down his chin under the oxygen mask and his tongue clinging to the roof of his mouth, with the crackling static sounding now like mocking laughter, and the intense concentration and the buffeting draining away his strength until his arms and legs were stiff and numb and his brain began to slip in and out of action like a faulty set of gears.

He managed to hold his height, but the fuel gauge fell dangerously low. He had no idea in which direction he was flying, nor even if he was keeping anything like a straight course. His only hope was that by sheer luck he would blunder out of the storm and pick up some landmark. And, with minutes to spare before his tanks dried, that is precisely what happened.

He came out over a low, rugged stretch of coast which he instantly recognized as South Cornwall. He went down to about 200 feet over the sea, a couple of hundred yards out from the rocks, and tried to find a flat place where he could get down. It was raining hard and very dark, but through the streaming windscreen he could just discern the outlines of the cliffs, and hills rising inland. At length he thought he saw a good stretch, on a headland. He switched on his powerful landing light and went in.

By now his fuel gauge was almost reading "empty" —there would be no time to make a proper inspection of the ground. The beam of the landing light, stabbing forward and downward from under the port wing through the driving rain, splashed on to what appeared a more or less level field of tufty grass. He could see no boulders or trees. He couldn't be sure whether the place was long enough to permit a normal landing, with the undercarriage lowered. But a belly-landing would damage the prop and the underside of the fuselage: there would be less shame attached to this episode if he could manage to bring his plane in undamaged. Besides, indefinable instinct told him that the grass stretched well away into the blackness ahead. So he put the wheels down.

He waffled in with the nose high and plenty of engine. Slamming the canopy back, he pulled his goggles down and thrust his head out into the deluge, peering along the side of the cowling at the arc of light leaping over the ground ahead of him, growing larger by the instant. The chill rain stung his cheeks, streamed down behind his oxygen mask and filled his mouth and nostrils, making him cough and gasp. Not until he'd eased the stick back to check the descent did he realize that the field sloped steeply upwards. He was landing uphill.

He gunned the engine hard and coaxed the nose up just in time. The Spitfire literally flew into the side of the hill, hitting the grass with a thump that jarred his bones and rattled his teeth. But the landing gear didn't collapse and he careened on, up the bumpy gradient, lurching and rocking, frantically stabbing the brakes on and off, fighting to bring her speed down. Just when he'd begun to think all would be well, the gradient decreased sharply, the plane bucked viciously and went over on to her nose. The prop blades splintered as they ploughed into the wet earth. The Spitfire seemed to stand on her spinner and for a dreadful instant he thought she was going to somersault, but she came to rest like that. With her tail in the air, and her nose in the mud.

He scrambled out and, not caring about the downpour, lay down full length and feasted his lungs on air that was sweet with the smell of the soaked grass. After a while he suddenly felt foolish lying there in the dark and the rain, so he got up, slowly and stiffly, and went to stand in the shelter of the fuselage. After several tries he managed to light a cigarette. By that time lanterns were bobbing through the darkness towards him.

They were land workers, and they told him he was a few miles from the town of Liskeard. Then they threw a coat about his drenched shoulders and led him towards a farmhouse. On the way they followed a line of high tension wires which ran across the field at the seaward end. The pylons were small, only about twenty-five feet tall. Midway between each pair of pylons the power lines, carrying thousands of volts, sagged to within twenty feet of the ground.

"Noice bit o' judgment that were, m'dear," one of the workers said. "Us were watchin' yer come in, for us seed yer loight from t'linney up thar, see? Us never reckoned as ye'd squeeze in under they woirs in t'dark that way, honest t'God us never! But yor done it a'right, noice as y'please. . . ."

Tuck just smiled and nodded. He didn't care to tell him the truth—that he hadn't seen the wires, that he'd passed under them by sheer good luck. He was very glad of the brandy they offered him when they reached the farm's snug kitchen.

Next morning, with a pound note borrowed from the Liskeard police inspector who'd put him up for the night in his home, he took a train back to Pembrey. Group Captain Hutchison, the station commander, gave him a mild rocket—"Bloody disgraceful, you know damned well you should have turned back to Northolt long before you got into the heart of that storm. . . ." —but there was no official disciplinary action. An R.A.F. crash crew dismantled Tuck's Spitfire in the Cornish field and brought it back by truck. All it needed was a new prop.

*   *   *

About this time the wonderful news reached him that Roger Bushell was a prisoner of war, unwounded. And—typical of the man!—somehow he'd managed to have a report of his May 23rd combat, in his own handwriting, smuggled through Underground channels in occupied Europe back to London! Air Ministry sent the squadron a copy of this:

"I was shot down by Messerschmitt 110s, but managed to get two of them first. As soon as the battle started about four or five of the Messerschmitts fell on me and, oh, boy, did I start dodging! My first I got with a full deflection shot underneath. He went down in a long glide with his port engine pouring smoke. I went into a spin as two others were firing at me from my aft quarter. I only did one turn of the spin and pulled out left and up. I then saw a Messerschmitt below me and trying to fire up at me, so I went head-on at him, and he came head-on at me. We were both firing, and everything was red flashes. I know I killed the pilot, because suddenly he pulled right up at me and missed me by inches. I went over the top of him and as I turned I saw him rear right up in a stall and go down with his engine smoking. I hadn't got long to watch, but he was out of control and half on his back. My engine was badly shot up and caught fire. My machine was pouring glycol. I don't quite know what happened, but I turned things off and was out of control for a while but got straight at about 5,000 feet.

"I shut everything off and the fire went out and I glided down. I landed just to the east of Boulogne and, of course, imagined I had come down in a friendly territory. The machine was blazing, but I had a look at it and could see some pretty hefty holes.

"I sat by my machine and when a motorbike came down the road I thought it was French. It wasn't, and there was nothing to be done about it. Thereafter I had a long journey here. This is an Air Force Camp, where we are treated very well indeed."

They held a terrific party that night at the *Stepney.* None of the recent arrivals, just the survivors of the old gang Roger had molded into a squadron and led into action for the first time—Tuck, Bartley, Havercroft, Howard Hill, Holland, Wright, Roy Mottram, Johnnie Bryson and Eddie Edwards. Maloney produced champagne, explaining: "I'm being let off light, gentlemen, for if this Mr. Bushell had come to Wales with you, I'm sure I'd have been down a dozen more bottles by now."

* * *

By mid-July the Luftwaffe had begun the real, round-the-clock assault. Composite patterns of bombers and fighters were crossing the Dover Strait every few hours. Their main objective was to smash Fighter Command—with their massive numbers to wear down the British squadrons by destroying the Spits and Hurris faster than they could be replaced. And they could count on at least two months of good flying weather in which to accomplish their task, and open the way for a land invasion.

At the time, of course, few of the operational pilots realized just how grave the situation was—that Britain was entering perhaps the most important period of her entire history, that a few hundred of them were all that stood between the nation and subjugation. They only knew that at last it had boiled down to a straightforward fight. No weighing of words or motives: they simply accepted the situation and cheerfully applied themselves to the hardest job in the world.

On July 26th, six Hurris of Caesar Hull's squadron, No. 43, without hesitation took on a bellowing herd of forty Dornier 17s and forty Messerschmitts. They downed five, possibly more, and broke up the big formation without loss. Throughout the command, morale rocketed—except on 92. They still languished in Wales—flying hard, but chasing only elusive recce. machines, sometimes getting in a quick squirt or two, but seemingly doomed always to lose their fast, stripped-down quarries in the puffy summer clouds that grazed, day after day, above the Carmarthenshire peaks and the Bristol Channel. Not until early August, when the fury of the conflict mounted suddenly, were they thrown into the main fight, and ordered east to patrol over the Channel and the southern counties. And from then on they were in the thick of it for long, violent weeks, caught up in the great sweep and rage of the biggest aerial battle in history.

9

When the first "flap" came through, Tuck was having a bath. Hearing the "scramble" bell pealing, ground crews shouting and engines firing, he leapt out of the water and struggled into his clothes without bothering to reach for a towel. But by the time he reached dispersal Kingcome had already taken off with seven others.

Titch Havercroft, Bobbie Holland and Sergeant Peter Eyles—one of the replacements who'd joined them after Dunkirk—were taxiing out, also having arrived late at their aircraft. He caught up with them as they climbed, formed them in loose echelon and then received a course from ground control which he realized, with a glow of excitement, would take them straight across England to the Sussex coast.

He wriggled about in his seat, uncomfortable because his clothes were sticking to his damp body. There was no sign of Kingcome and the others, but though he cursed at missing them, his lateness proved a blessing. For the main body of the squadron returned without having sighted an enemy machine, while the smaller section, stooging around off Portsmouth looking for their colleagues, caught three Ju. 88s speeding out from the land.

The raiders had dropped their dirt and were going all out for home, a few feet over the smooth water. Tuck wheeled and sliced down after them.

"Holland and Eyles were a little slow in turning, but Titch stayed right up with me. The 88s were in a fairly wide line abreast, and honestly you couldn't see anything between them and their own shadows on the surface. I managed to get up behind the port one and hit him hard. He started to lose speed immediately and streamed black oil and muck. I gave him another bash. He went *splat!* into the water, and as I flashed over him I could see him ploughing along like a bloody great speedboat in the middle of a tremendous cloud of white spray.

"All this time Titch was banging away at the starboard one. I tried to get on to the leader, but by now we'd lost the extra speed from our dive, and it was all we could do to keep up. The Ju. 88 was a wonderfully fast kite, especially when it had unloaded and the pilot was homeward bound with a Spit up his backside. . . .

"I was at long range—I think about 900 yards—but I was managing to lob a bit on to him. This was one of the many times I cursed because I didn't have

cannon. I was hitting him all right, but nothing was happening. We got well out over the Channel, and I remembered to take a quick check on fuel—we'd been bending our throttles on the end of the "emergency" slots for minutes on end now. My gauge was reading a bit low, so I lined up very carefully and gave him a last, long burst. This time a few bits flew off him. Then I called up Titch and we broke off the attack. Titch left his Ju. streaming a thick trail of oil.

"Heading back for the land we saw Holland and Eyles, very low on the water, circling the wreckage of the one I'd shot down. The crew of three were huddled in their rubber dinghy, looking up at the Spits, obviously very worried. I think the poor sods were afraid we'd strafe them.

"I climbed to fifteen hundred, called up base and let them get a good fix on the position so that the Air Sea Rescue boys could come out and collect. Then all four of us went down and once around the dinghy, making V-signs and rude versions of the Nazi salute.

"Brian Kingcome was furious when we got back. His crowd hadn't seen a thing. I told him if only he'd take a bath more often he'd be more successful in life."

The very next day Tuck was leader of a section which intercepted another three Ju. 88s at 15,000 feet eight miles north of Cardiff.

"This was one of the most interesting combats I ever had. There was Titch, myself and a new lad whose name I can't for the life of me remember now. We were directed on to the Huns by ground control. We were heading almost exactly north, above some flat stratus cloud when control said: 'Your plots are coinciding now. Bandits heading due south. Can you see them?'

"Sure enough, within seconds Titch spotted them—three black dots coming smack at us. The closing speed was terrific, so I decided there wasn't time for a head-on attack, although ordinarily I was a firm believer in that method—the Jerries had no armored plate in the front, and if you hit the nose or cabin you had a very good chance of knocking out the whole crew, who always liked to sit close together and hold hands; not like our bomber chaps, who were spread out in tail and nose turrets, in contact with one another only by intercom.

"I pulled up a bit and took the boys round in a very steep turn, so that we came out on their tails well within range just a few seconds after they'd whipped by under us. Their rear-gunners let fly and suddenly I saw Titch give a lurch, and then drop back, spewing glycol. He'd got a couple right through the radiator.

"He called up, swearing like a drunken trooper, so I knew he wasn't hurt. Not surprising—he was such a small target they could never hope to hit him. . . . He'd bags of altitude so although there were big hills below us, he'd a good chance of getting her down somewhere. The youngster and I pressed on.

"I was justa bit worried about this kid—he was very inexperienced. These rear-gunners were good shots and wily enough to work together, all three bashing at one of us at a time to get a wicked crossfire. After a bit I decided to pack in this stern attack and try something else.

"We broke away, climbed a little and screamed ahead on full throttle. These particular 88s weren't nearly so fast as some I'd come across—or maybe it was just that at this height the Spit had the measure of them. Anyway, we got ahead quite easily, turned around and got set to have a go at them head-on.

"By now I suppose we must have been right over Cardiff. As they came at us we throttled back hard and put our props in fine pitch to brake us—to give us the lowest possible closing speed. We got everything lined up, trimmed up and set nicely long before they were in range. I got the dot of my sight resting neatly above the leader's canopy, and I yelled at the kid to do the same. Then we just waited for them. With that closing speed I opened fire well out of range— probably it was around 2,000 yards. This would give me a chance to correct once I saw where my tracer was going. But I was lucky, and didn't need to correct, because from the very start of that long burst I seemed to be on the mark all right. I held the dot a shade high for a second or so, and then at the last instant dropped it full on to his canopy.

"By the time we'd pulled up and turned around, the leader had dropped right out and was going straight on down. Afterwards we learned he'd crashed on the outskirts of Cardiff. They got a couple out of the wreckage alive.

"We tore ahead of the remaining two, and repeated the same performance exactly, out over the Bristol Channel. Would you believe it, the stupid clots still stuck to the same course! They could have ducked down into the cloud, a few thousand feet below, and we'd have had very little hope of finding them again. Extraordinary, those Germans—they'd probably had orders from the information leader to maintain this height, speed and course, and even though he'd gone now, they were still obeying! Quite a lot of them were like that—'goose-step' pilots.

"On the second head-on attack I knocked some big chunks off the port machine, and he went over in a sickening roll. He finished up on the south bank of the Channel.

"The youngster had got some hit on the third one, and now, left alone, at last this chap forgot orders, stuck his nose down and tucked himself up in the stratus. Both of us managed to give him a farewell burst just before he disappeared—like a whale sounding. He seemed to have slowed a lot, and he was definitely streaming oil and glycol. I felt sure he couldn't live, but when we got back the Intelligence types couldn't credit us with a definite 'kill.' An 88 had crashed well to the south, a little way inland, but some Clever Dick on another squadron claimed it.

"Titch had made a very nice belly landing on a mountainside near Aberdare, then was nearly done in by the local Home Guard who took him for a squarehead. They marched him along at bayonet point for a while, but let him go eventually because they realized no German could know so many weird combinations of British cuss words. But we couldn't scrounge any of Maloney's bubbly that evening, because only an hour or so before dark I was scrambled again with Allan, Bobbie and Peter Eyles—and if we'd known what we were in for we'd have taken our toothbrushes and pajamas. . . .

"More or less over the Swansea dock area we nabbed

two more 88s. It was gray and drizzling now, and the light was pretty poor. The Huns were steaming for home and when they spotted us they nipped in and out of every patch of cloud. We chased them out over the Bristol Channel, taking a quick poke at them every time they emerged in the clear for even a second or two. But there was no chance to damage them seriously. Here again I'm convinced cannon would have done the job.

"In the end the light beat us and we lost them completely. We were now over South Cornwall—that wild stretch of coast that reminded me very forcibly of my recent misadventure in the storm, especially since we'd been airborne for two hours now and our tanks were uncomfortably light.

"We went down quite low and came snooping up the coast, trying to remember what fighter 'dromes there were in the area. The weather worsened very rapidly —it began to rain heavily and in no time it was almost pitch dark. We went on and on, straining our eyes and calling on the r/t, but never a sign of a 'drome.

"Then suddenly, right on the edge of some nasty looking cliffs, I saw something fluttering—a tiny windsock. It turned out to be a satellite field, very small, which I managed to identify as Cleave. It was meant for communications aircraft and trainers. It seemed deserted.

"Between the top of the cliffs and the base of the cloud there was only about two hundred feet. It was going to be a very tricky business—you'd have needed a shoehorn to get a Spit in there even in perfect conditions."

The wind was coming from the water—up over the cliffs. That meant they'd have to go in from the landward side, so if anybody overshot he'd finish up plastered over the rocks below. Tuck called up on every channel for permission to land and maybe some flares, but there wasn't a hint of life down there. Not even a light, or a Very cartridge.

He told the others: "I know you're short of gravy, but don't attempt a landing until you see how I get on. If I make a mess of it and go over the edge, come in with your wheels up." By waffling in on the prop and braking like mad, he got down all right, and finished up about twenty feet from the big drop. He taxied back to the downwind boundary, called them up again and said: "It's just all right, but bring them in as slow as you can, with bags of throttle, and drop them right over the hedge. Switch on your navigation lights—I'll watch your approaches and let you know how you're doing. You first, Bobbie."

It was young Eyles he was worried about. The sergeant was the least experienced, and his voice had been distinctly shaky as he acknowledged Tuck's orders. Best leave him till last, so that he could watch how Bobbie and Allan did it.

They did it with even less to spare than their leader. By that time the sweat was lashing down Tuck's face as he sat in his darkened cockpit, with the rain drumming on the canopy, staring up into the blackness at the faint, rocking lights of the last Spitfire.

Suddenly Eyles' voice exploded in the earphones, high and breathy: "My gauge reads zero! For Christ's sake get me down!"

"Easy, Peter! Do exactly as I tell you. No need to get panicky!" Tuck talked slowly, very carefully. In that flat, impersonal voice. Talked almost without pause, without the slightest change of speed, even when the N.C.O. misjudged his first approach and came in much too fast.

"Round again, round again! You're going to overshoot."

But Eyles was in the spell of that empty fuel gauge— afraid that he hadn't enough gas to make another circuit. So he didn't respond immediately to Tuck's order, but hesitated, wheels still groping downwards for the ground, wings rocking drunkenly, hanging on his prop there in the rain and the blackness almost halfway across the tiny field and more than halfway to an ugly death. Still Tuck didn't raise his voice, but somehow he spoke with increased authority.

"Sergeant, I'm telling you to go round again."

A few more seconds of suspense, then Eyles' engine swelled to full voice and he rocked back up into the night. Tuck, Holland and Wright sat in their cockpits with their jaw muscles knotting and their lungs frozen, watching the boy's lights going round, dreading the sudden splutter and cough which the Merlin would give as the carburetor dried. Then there would be a short, agonizing silence . . . and suddenly the grinding, splintering crash somewhere out there in the night. . . .

Tuck kept talking at the same pace, every word almost on the same pitch. "This time get your speed well down, bring her in right on the stall . . . let her waffle, you can wait till you feel her wanting to drop . . . bang on the throttle if she starts to fall out of your hands.

"All right, bit more throttle . . . check, check—back still, that's it . . . just hold her there . . . keep your wings straight . . . you're bang on now, just drag her along . . . drag her along . . . stand by to cut your throttle . . . stand by . . . *cut!*"

Eyles touched down in the first five yards of the field and stopped shorter than any of the others. In fact, his was the best landing of all.

"Good show, Peter," that clipped voice said. "Now let's see about supper."

Two sleepy, drippy airmen in ungainly rubber capes and gumboots appeared out of the blackness with torches, and directed them to dispersal. But they had to fetch a tractor for Eyles' machine. It hadn't enough fuel left to taxi the 400 yards.

* * *

At Hornchurch King George VI invested him with the D.F.C. Just five minutes before the monach's car arrived, with the whole parade drawn up in perfect order, somebody noticed that Tuck had neglected to sew on to the left breast of his tunic the tiny metal hook on which the cross could be hung.

He sprinted off to the Orderly Room, where the staff were working as usual, and explained his predicament. A W.A.A.F. typist scurried to the washroom and reappeared, blushing, with a small hook snipped from her undies. It wasn't the recommended size or color, but it would serve. Somebody conjured up a needle and thread, but the little W.A.A.F.'s hands shook so much with excitement that the job took all of three minutes. He got back on parade, breathless and perspiring, just as the Royal car swung on to the tarmac.

Several officers and N.C.O.s were decorated that day: the man standing on Tuck's right was Sailor Malan. The old friends had only a brief talk afterwards —there was no time to celebrate, for each had to get back to his squadron.

*   *   *

The Vickers Aircraft Company had taken over Brooklands, the prewar motor racing track, as a test field and experimental plant. Tuck was sent there one day to have a new camera-gun fixture installed in his aircraft. The job done, he hopped over to Northolt to have lunch with Group Captain "Tiny" Vass, an officer he'd known since training days. In the middle of the meal the alarm went and the squadrons scrambled. Vass grabbed a phone and learned that a big battle was developing off Beachy Head, Sussex. At that moment the air-raid warning sounded. Station orders were that all nonflying personnel were to take shelter in slit-trenches during raids.

Vass, a mountainous man with the strength of a farmhouse, grabbed Tuck's arm and propelled him out of the mess.

"Come along, Tommy. Down the bloody bunker!"

Protesting, Tuck was shoved into a damp hole where an N.C.O. unceremoniously rammed a tin hat on his head. Looking out across the grass he could see his aircraft—the only one left at dispersal. He started to clamber out but Vass grabbed his ankle and hauled him back. They argued for precious minutes before Vass finally shrugged and said: "Dammit, I can't give you permission to take off, not on your own. But . . . well, I can't be everywhere, I can't see everything that goes on, can I?" He turned his back and concentrated on surveying the sky to the east. Tuck went west—scrambling out of the trench and pelting across to his machine. He started it himself and was airborne within two minutes.

He picked up Hornchurch control and followed their directions. As he crossed the coast, far to the north and very high he could see a tremendous tangled mass of vapor trails, and tiny glints of metal . . . then the hurtling, weaving fighters, mere dots, flitting in and out of sight, like dust motes in the sun's rays. He advanced his throttle through the gate into "emergency power" and clambered up frantically to join in the fray.

Soon he could distinguish 109s, Hurris, Spits, 110s, a few Ju. 88s. . . . It was the biggest fight he'd yet seen, an awe-inspiring spectacle that made his throat tighten and produced an odd, damp feeling at the temples and wrists.

Then suddenly, far below him two 88s passed, very close together, striking out for home at sea level. He turned out from the land, away from the main scrap, and with a long, shallow dive got well ahead of them. Then he turned again, due west, dropped low over the water and made a head-on attack.

The port one reared up so violently as the Spitfire's bullets ripped into the cabin that its slender fuselage seemed to bend backwards, like the body of a leaping fish. Then one wing dipped. The plane cartwheeled and vanished in an explosion of white water. It was exactly as if a depth charge had gone off.

He pulled up sharply into a half loop, rolled off the top and dived hard after the second bomber. He passed above it, raced ahead, and came round again for another head-on attack.

Tracer came lobbing leisurely at him: from this angle it wasn't in streaks, but in separate, round blobs, like a long curving chain of electric light bulbs. The stuff was strangely beautiful, the way it glowed even in the broad daylight. At first it was well out of range, but as the two aircraft raced towards each other, suddenly he was flying in the broad jet of the Junkers' forward guns, and the tracer seemed to come alive and spurt straight for his face at bewildering speed. He concentrated on his own shooting, and saw his stuff landing on the bomber's nose and canopy.

The enemy's silhouette, limned almost black against the sunlight, remained squarely in his sight, growing with incredible rapidity, and all the time those wicked blue flashes twinkled merrily on it. It came on and on like that, calm and beautiful and stately as a giant albatross, straight and unflinching as though it were some purely automatic missile, an unfeeling super arrow, scientifically, inexorably aimed so as to drive its point between his eyes.

He had the sudden unsettling conviction that this one was different from all the others. This one was more dangerous. It wasn't going to stop firing at him, it wasn't going to break off no matter how much lead he pumped into it.

This one could be death.

All this was happening, all these thoughts and feelings were crowding on him, in the space of a mere two or three seconds. But everything was so clear, so sharply focused. The moment seemed to stand still, in order to impress its every detail on his mind.

The silhouette grew and grew until it seemed to fill the world. He clenched his teeth and kept firing to the last instant—and to the instant beyond the last. To the instant when he knew they were going to crash, that each had called the other's bluff, that they could not avoid the final terrible union.

Then it was a purely animal reflex that took command, yanked the stick over and lashed out at the rudder. Somehow the Spitfire turned away and scraped over the bomber's starboard wing. There could have been only a matter of inches to spare, a particle of time too tiny to measure. Yet in that fleeting trice, as he banked and climbed, showing his belly to his foe, several shells smashed into the throat of the cowling and stopped up the Spitfire's breath. The elaborate systems of pipes and pumps and valves and containers which held the coolant and the oil, and perhaps the oil sump too, were bent and kneaded into a shapeless, clogging mass that sent almost every instrument on the panel spinning and made the Merlin scream in agony.

"With what speed I had left I managed to pull up to around fifteen hundred feet. I was only about sixteen miles out, but I felt sure I'd never get back to the coast.

"I can't understand why that engine didn't pack up completely, there and then. Somehow it kept grinding away. I was very surprised, and deeply grateful for every second it gave me.

"As I coddled her round towards home I glimpsed the 88 skimming the waves away to port, streaming a lot of muck. In fact, he was leaving an oily trail on the water behind him. I had the consolation of thinking the chances were that he wouldn't make it either."

At the time, Tuck was heartily glad to see the Junkers in as poor shape as himself, and he hoped fervently that the German would crash into the sea.

"I trimmed up and the controls seemed quite all right. The windscreen was black with oil. Temperatures were up round the clocks and pressures had dropped to practically zero. But she kept on flying after a fashion. Every turn of the prop was an unexpected windfall—that engine should have seized up, solid, long before this.

"I knew it couldn't last, of course, and I decided I'd have to bail out into the Channel. It wasn't a very pleasant prospect. Ever since my prewar air collision I'd had a definite prejudice against parachutes. But the only alternative was to try to ditch her, and a Spit was notoriously allergic to landing on water—the air scoop usually caught a wave and then she would plunge straight to the bottom, or else the tail would smack the water and bounce back up hard and send you over in a somersault. Bailing out seemed the lesser of two evils, so I opened my hood, undid my straps and disconnected everything except my r/t lead.

"It got pretty hot about now. The cockpit was full of glycol fumes and the stink of burning rubber and white-hot metal, and I vomited a lot. I began to worry about her blowing up. But there were no flames yet, and somehow she kept dragging herself on through the sky, so I stayed put and kept blessing the Rolls-Royce engineers who'd produced an engine with stamina like this. And in no time at all I was passing over Beachy Head.

"I began to think after all I might make one of the airfields. The very next moment, a deep, dull roar like a blowlamp started down under my feet and up she went in flame and smoke.

"As I snatched the r/t lead away and heaved myself up to go over the side there was a bang and a hiss and a clump of hot, black oil hit me full in the face. Luckily I had my goggles down, but I got some in my mouth and nose and it knocked me right back into the seat, spluttering and gasping. It took me a little while to spit the stuff out and wipe the worst of it off my goggles, and by that time I was down to well under a thousand. If I didn't get out but quick, my 'chute wouldn't open in time.

"It wasn't the recommended method of abandoning aircraft—I just grabbed one side with both hands, hauled myself up and over, and pitched out, head first. As soon as I knew my feet were clear I pulled the ripcord. It seemed to open almost immediately. The oil had formed a film over my goggles again and I couldn't see a thing. I pushed the goggles up, then it got in my eyes. I was still rubbing them when I hit the ground."

It was an awkward fall and he wrenched a leg and was severely winded. He was in a field just outside the boundaries of Plovers, the lovely, old world estate of Lord Cornwallis at Horsmonden, Kent, and several people had witnessed his spectacular arrival. The blazing Spitfire crashed a few hundred yards away in open country. An estate wagon took him to the house, where His Lordship had already prepared a bed and called his personal physician. But Tuck, once he'd stopped vomiting, insisted on getting up to telephone his base —and once on his feet, wouldn't lie down again. He had a bath, leaving a thick coat of oil on His Lordship's tub, then despite the doctor's protests, borrowed a stick and hobbled downstairs in time to join the family for tea.

But after that, very suddenly, exhaustion took him. They helped him back upstairs and he slept deeply for three hours. When he awoke his leg felt better and his host's son, Fiennes Cornwallis, was waiting to drive him to Biggin Hill, where a spare Spitfire would be available.

"Drop in for a bath any time, m'boy," said his Lordship.

At Biggin Hill the only accommodation they could offer him for the night was in the station's hospital. The patient in the next bed was the pilot of a Ju. 88 shot down a few miles from the airdrome that same afternoon. He was very young and very husky, with a shock of wheat-colored hair, a bright pink complexion and rather dreamy blue eyes. His injuries weren't serious: minor burns on hands and arms, bruised ribs and a cut head. But he'd been badly shaken; all the stiff, Germanic pride had been knocked out of him and he looked rather lost and sorry for himself, like an overgrown schoolboy whose latest prank had misfired and landed him in adult-sized trouble.

Tuck gave him a cigarette. They found they could converse quite well, for the lad had some English, and Tuck's German was passable. After a while Tuck fell silent, thinking how dreadful it must be to be a prisoner of war, cooped up behind barbed wire in an alien land. . . . Then he noticed that the German, though in hospital-issue nightshirt, was still wearing an Iron Cross, slung on a broad ribbon about his neck.

"I see you've been decorated."

"Yes. Iron Cross, second class. And that is the ribbon of the Distinguished Flying Cross that I see on your tunic, is it not, Squadron Leader?"

"Right."

"Why do the British wear only the ribbon? Why not the cross too?"

"I don't know—this is the custom, that's all. Except on very special occasions."

The German accepted another cigarette. When Tuck had lit it for him he asked: "How many victories have you?"

Tuck didn't like this at all—it was the one question he'd taken for granted would be left out of this unique discussion. It seemed to him almost indecent. His voice was clipped and cold. "I'm credited with eleven aircraft destroyed." He purposely left out the victory he'd added that very day, and several others which hadn't been confirmed and made "official." He was wishing to hell he'd never started all this. . . .

The German pondered Tuck's answer for many minutes in silence, sitting up very straight in the bed, watching him intently, trying to decide if he was telling the truth. Then he began a long series of questions which Tuck answered curtly at first, and then with growing warmth—because they were good questions, from another war pilot, about the calling which had taken up both their lives. Soon he forgot his earlier embarrassment and talked as freely and excitedly as he would in his own mess to one of the young replacements.

At the end of it, the German did something quite startling. Very solemnly, he took the Iron Cross from his neck and handed it to the Englishman.

"Squadron Leader . . . I should like you to have this. For me the war is finished, but somehow I think you have a long way to go yet. It would be very nice for me to know, when I am locked away in the prison camp, that my cross is still flying—still free, as it were. So please take it with you every time you go up. I hope it will bring you luck."

Tuck didn't want the medal, but the boy was adamant. So, for once at a loss for words, he accepted it, and promised to carry it always.

* * *

Precisely one week after his parachute escape he was in trouble again. And lucky again.

He had with him Bobbie Holland and Roy Mottram. An unidentified aircraft had been reported off Swansea. A day or two earlier the big oil storage vats at Pembroke Dock had been hit, and they were still burning fiercely. The great pall of smoke reminded him of that first day's fighting over Dunkirk. Between 3,000 and 4,000 feet there was a solid shelf of white cloud, and running through this was a distinct, oily black ribbon.

Control told them a small coaster coming up the Bristol Channel was being bombed by a Do. 17, and gave them a course to steer. They dived below the cloud and found the ship, but couldn't see the raider. All at once a plume of spray sprouted just off the coaster's port bow. Still they couldn't spot the Dornier, but now at least they knew its approximate position and course. At full throttle they flashed over the ship and climbed through the cloud. The topside was smooth as a billiard table. Still nothing. Tuck called to the others to stay up top, on lookout, then nipped back underneath.

As he broke cloud and circled, he spotted the big, logger-headed Dornier making another run on the coaster. Another white plume blossomed close by the little vessel. On the deck he could see a group of seamen fighting back gamely with a couple of ancient Lewis guns.

The Dornier saw the Spit curving in at him, and quickly pulled up, into the cloud. Tuck followed him in, overtaking very fast. As the blinding whiteness struck the windscreen he throttled back hard—to stay behind him, with luck to catch the vaguest outline of his tail.

The best part of a minute passed. His straining eyes saw nothing. Then came a series of deafening thuds and the Spitfire kicked and leapt like a startled foal.

Christ, he was being clobbered . . . the Hun's rear-gunner must have X ray eyes! He kicked his rudder savagely, yawing about violently in an effort to get out of the fire, but couldn't. More thuds, more jolts and shuddering. He narrowed his eyes to slits and shoved his face forward, close to the windscreen. If it could see him, then he ought to see it! But—only the whiteness.

Then his port wing lifted joltingly and, glancing out, he noticed a couple of holes in it. He skidded to the right and at once saw, immediately below and very slightly ahead, the shadow outline of the bomber. He'd been sitting almost on top of it! He closed the throttle, dropped down and slid in directly behind it, ignoring the rear-gunner's furious blasts. Then he opened up the engine and edged up on it. From what couldn't have been more than fifty yards he dealt it a long, steady burst. Then he rose a little to one side, and with his guns roaring again gently brought his nose down slantwise and literally sawed right across it, from starboard engine through the fuselage to port engine and out to the wing tip. The cloud thinned suddenly, and he could see holes as big as his fist appearing all over it. But at this low speed the recoil of the second burst threw him into a stall, and he went hurtling down into the clear air.

His engine was critically damaged, spluttering and rasping and leaking glycol. But as he regained control and brought the nose up he saw the Dornier plunge vertically into the water, less than half a mile from the little ship it had failed to hit.

He called up Bobbie and Roy and told them the score. Bobbie had gone steaming off somewhere after a shadow, but Roy came swooping down, slid in alongside, and took a good look.

"One *helluva* mess underneath," he reported—and he spoke with a chuckle in his voice, as if his leader were a schoolboy who'd just driven a cricket ball through the vicar's greenhouse. "I say, you're in *beastly* trouble and no mistake!"

The smashed engine was losing power fast. Tuck decided he was very unlikely to make the shore. It would have to be the 'chute again. Angrily he made ready to depart, sliding back the canopy, undoing harness and oxygen pipe.

"Going somewhere?" Mottram inquired.

Tuck reached out a hand to disconnect the r/t lead, but surprisingly, the engine began to provide spasmodic bursts of power. He hesitated. The shoreline drew closer. Then a last, gallant pop or two and the Merlin died. Dead ahead there were sheer cliffs with flat, browned grassland on top—St. Gowan's Head. He had about twelve hundred feet. Only seconds to make the decision, and his judgment had better be precise.

He decided he could just make it. He could stretch his glide and set her down on the clifftops. The ground looked rough: he'd have to make it a belly-landing.

He held the speed at a hair's breadth above the stall mark and, forcing himself to relax, to be delicate and to "feel" every inclination of his aircraft, settled down once more to fly for his life.

"Ha!—*now* you've had it!" cried Mottram after they'd descended another two or three hundred feet. "Should've gone while you had the chance. You'll *never* get over these cliffs!"

Tuck shot him a hateful glance. Mottram, no respecter of rank, responded with a hoot of laughter and some rapid-fire V-signs. Predicting disaster was his way of giving comfort.

Right up to the last moment the issue was in gravest doubt. Mottram, irrepressibly pessimistic, stayed right on Tuck's wingtip all the way, until the crippled Spitfire, fluttering in the first tremors of a stall, grazed the brink of the cliffs and bounced on its belly on sun-cracked, rutted ground. Then he yelled: "Jolly good!" and got out of the way.

Only as he wrestled to set her down again did Tuck remember, with sickening apprehension, that he'd undone his safety straps. No time to rectify that now—he could only hope he'd get away with black eyes and a broken nose, not crack his skull. . . .

Then the tail struck a bump and the Spitfire bounced up again, this time for all of a hundred feet. All hope drained away. Now he would surely stall—a wing would drop, she'd plow in on her back. . . .

She seemed to stop in midair and drop vertically, like a lift, but shaking herself like a dog after a swim. He shoved the stick hard forward and worked the rudder feverishly to try to keep her straight.

Steel bands around his chest, thorax throbbing painfully, damp-faced and dry-mouthed. He was like a man lashed face upwards on the guillotine—in another instant he would be watching death hurtling at him. But though the needle was for several seconds distinctly below stalling speed, the Spit didn't drop a wing. A creature of true breeding, she kept her poise— and suddenly she was responding, she was flying again, she was gliding down smoothly and steadily!

This time he was able to flatten out with a shade more speed, and consequently more control. Out of the corner of his eye he saw Mottram away to the left, very low, going round in a steep bank, watching him. Then he spotted a hedge across his path just ahead and thought: "That's handy, that'll break my fall a bit, I'll try to touch down right on it."

He judged it perfectly. But instead of passing through the leafy barrier, the Spitfire stopped dead, as if she had hit a wall.

She *had* hit a wall—a dry-stone dyke hidden by the hawthorn. She came from eighty miles an hour to a standstill in about five feet, but Tuck went right on travelling, out of the seat—luckily not upwards, not out of the cockpit, but horizontally, in the general direction of the instrument panel, on and on into an ocean of darkness.

When he awoke he could remember nothing at first and couldn't think where he was. No pain, no sound at all. Something was pressing down on his head, though, and there was a strange piece of metal wound around his leg. Odder still, he seemed to be all rolled up in a ball.

He stared at the twisted piece of metal and recognized it as the control column. He looked around some more and found he was sitting on the rudder pedals. Then he knew where he was: he was stuffed into the small space under the instrument panel.

There came to his ears, gradually, a faint hissing and dripping, and he caught a whiff of gas. That made him wriggle out very quickly. He walked about ten paces and sat down with a thump. It was a great effort to raise an arm and wave to Mottram's low-circling machine, but he managed it. Then he decided to have a nice, quiet snooze.

In hospital that night he persuaded a young nursing sister to let him go to the telephone in her office. He was talking to "Mac," the adjutant at Pembrey, when suddenly he was greatly surprised to find that he couldn't see anything, not even his hand holding the phone. He just managed to slur out: "Well, g'bye, ol' boy, g'bye" and fumble the receiver back on its hook. Then he did a forward somersault into oblivion.

Post accident shock, they called it, and they kept him on sedatives for two days. When finally he was discharged he suffered another shock: he had been posted from 92 to 257.

And 257 was a *Hurricane* squadron!

10

It was while he was packing his kit for the move to his new command that they told him Caesar Hull had bought it. They'd kept it from him while he was in hospital.

"Little Caesar," the small giant, the gentle tough. Gone. That hoarse, sweet voice stilled, that battered but beautiful face forever hidden. It was like a steel shutter falling, cutting off a whole region of the past, making a thousand dear memories suddenly remote, unreal.

Since Grantham they'd met only occasionally—by chance on "rendezvous" airfields, by arrangement in London for a few brief but boisterous weekend leaves. During the intervening periods Fighter Command's mysterious but efficient "barroom telegraph" had kept them informed of each other's doings. Somehow, despite the long partings—for instance, when Caesar was serving in Norway—their friendship hadn't faded, but flourished. They never wrote letters, yet they were never out of touch. After months without meeting, they had only to be face to face with one another for an instant and they were right back in the old, easy intimacy which had begun in the first days of their cadetship.

The last time they'd met, Caesar was celebrating a recent victory, in the mess at Tangmere. He and Peter Townsend had performed their famous "La Cachita" dance to a gramophone record. This began as a cross between a rhumba and a Maori *haka*, but developed quickly into a sort of combination wrestling match and steeplechase. They threw each other all over the room, turned somersaults, clambered over the chairs and tables, all the time waggling their bottoms in time with the Latin rhythm.

Later Caesar and Tuck had driven off on their own to a country pub, beaten the "regulars" at darts several times over and won many pints of bitter. A grand night; "one for the logbook." A fine, last memory to treasure. . . .

Throughout his entire air force career only three men became, in the full sense, his close friends. Caesar Hull was the first of these. Tuck kept his grief locked inside him and very soon—employing that remarkable gift for shutting out of his thoughts anything that could conceivably affect his work or his standing—he contrived to file the pain away in his subconscious repository. But first he had to know all the facts about his friend's death.

Caesar, he learned, had been leader of a group of about ten 43 Squadron Hurricanes which had tangled with a force of seventy-five bombers and around one hundred and fifty escorting 109s over the Thames Estuary. Impossible odds.

One by one the Hurricanes ran out of ammunition, and several were shot down because they continued to dive through the enemy formations in an attempt to break them up before they could get through to London —or at least to hold them off till other squadrons arrived.

Caesar's number two was Dickie Reynell, an Australian who was actually a civilian at the time—one of the top test pilots of the Hawker Aircraft Company. Air Ministry had recently agreed to allow a select number of test pilots to fly for short periods with operational squadrons, so that they might gain first hand knowledge of the service pilots' problems.

Poor Reynell had learned their problems all too well: he got separated from the rest of the squadron and was "bounced" by a flock of 109s. Caesar went to his aid, but his guns were empty. Both were shot down.

Reynell bailed out, but was apparently killed during his parachute descent. Caesar died in his cockpit.

Tuck was off on his own for an hour or two, absorbing the shock and adjusting his mind. He seemed his usual, carefree self when he joined Kingcome and the others in the mess for a farewell beer party. It was a rushed affair. His orders were to take up his new post immediately—that same evening. This note of urgency made him suspect trouble, so he was prepared for the worst when he took leave of his tried and trusty band and set off to join 257.

They were a sorry looking lot.

Scruffy, listless and leaderless, they were quarreling among themselves over trivialities, drinking hard but entirely without zest. Over the last few weeks they'd taken a severe mauling—this was probably the only squadron in Fighter Command to lose more aircraft than they'd shot down.

For all their failures and frustrations they blamed "organized chaos" up top—they really believed they were being betrayed by their leaders, that Air Ministry, Fighter Command and Group Headquarters were peopled by bungling old "Blimps" who didn't have a clue about modern airfighting and who were fast losing the war.

When Tuck walked into the 257 officers' mess at Martlesham they shuffled to their feet and nailed him with phlegmatic stares. To them he seemed a lean, mean and vainglorious person. He had a dry, arrogant face and he walked with his head high, like a blind man. The immaculate uniform, the glossy hair and the "Caesar Romero" mustache kindled their instinctive scorn for all forms of bullshine. (It was true that outwardly, nowadays he often seemed a noisy show-off and a bit of a fop, too. Lately, he had taken to using a long, slim cigarette holder, and some of his mannerisms were haughty indeed!)

Tuck didn't speak to them right away. He stood in the middle of the room and raked the sullen faces with a long, searching glance. Then he nodded to them to sit and walked hurriedly to the bar. He was accompanied by the squadron Intelligence Officer, Jeff Myers. Myers—small, plump, fortyish—had been a foreign correspondent for a big national daily before the war. He was a placid, soft voiced and gentle eyed Jewish intellectual, and he more than anyone else felt the hostile tension and knew how crucial the next few minutes would be.

Tuck ordered two pints of bitter. Hardly a flicker of talk in all that large room.

"Cheers," Jeff said huskily.

"Jolly good luck!" The voice was relaxed, almost flippant. Tuck downed most of his pint at a single hoist, released an appreciative sigh and then grinned at Myers. The grin seemed to alter his whole personality —you could almost feel the air grow warmer.

"Needed this, by God!" He held the glass up and squinted at it. "Nice drop of stuff, too."

"Not bad at all." Jeff stared at the mug in his hand. He couldn't think what it was doing there. He disliked beer, never accepted a pintful—at a push he could make half a pint, forced upon him, last all night.

Nobody joined them. They just sat watching, some pretending to read magazines, a few now beginning to murmur indistinctly. The adjutant didn't show up. Jeff wondered whether he should offer to perform the introductions, but decided to wait until Tuck gave him a clear hint.

Tuck was in no hurry. In the next thirty minutes or so he drank four pints. Jeff, wholly bemused, finished one and sipped his way well down into a second. They talked about the pubs around Martlesham, about London restaurants and cocktail bars, secondhand cars, a recent raid on Dover and the comparative merits of various B.B.C. comedy programs. Part of the time Jeff wasn't really listening: he was thinking: "Either this man is a gifted actor, or a plain bloody clot—or else, damnitall, he's completely at his ease!"

Jeff didn't notice the first of them drift over. He only knew that suddenly they were shaking hands with Tuck, looking a bit sheepish, and telling him their names, and he was buying them drinks. And then, in

no time at all, he had one or two of them talking—
cautiously at first, and then with amazing candor—
grinding out their grievances, the whole dismal history
of disastrous patrols, misdirections, confused tactics,
muffed chances and growing mistrust. In their eager-
ness to justify themselves they laid bare their own
blind bitterness.

Tuck didn't sympathize, but he didn't argue either. He listened, nodding, asking an occasional question, leaning back against the bar with his long legs
crossed and his long, narow eyes swinging smoothly
from face to face. Most of the time he was fingering
an oddly misshapen penny. Gradually, almost im-
perceptibly, he began to take control of the talk. His
questions became more frequent, more searching, and
then one or two of them weren't really questions at
all, but comments or suggestions.

Most of the pilots remained silent, suspicious, and
some openly hostile. But Myers was amazed, and deeply stirred, to see other faces losing their blank,
wrenched expressions, the eyes coming alive for the
first time in weeks. Watching Tuck from the edge of
the group, he decided that this bloke had the God-
given talent of succeeding with men, of winning over
some natural enemies, perhaps of inspiring them to
great effort. They didn't necessarily all come to like
him, but that was beside the point.

Jeff had the writer's habit of assessing and analyzing
character, often on the shortest acquaintance. He
tried now to make a preliminary summing up of Tuck:

A brilliant flying record—fourteen official "kills"
(but probably seventeen or eighteen was nearer the
truth) . . . fantastic luck and a series of incredible
escapes . . . reputedly a precise craftsman and a
phenomenal shot . . . a big drinker and party man, a
tireless, driving character who could go for days without sleep. . . .

Such was the growing legend, but Myers had
knocked around the world enough to know that no
man, no matter how gifted or successful, was without
human failings, and he tried to decide what Tuck's
faults might be. He guessed—and he wasn't far wrong

—that intolerance of others' weaknesses might be one, snap judgment another.

Tuck was, in fact, by now a very hard case. The death of Caesar Hull had made him more tense and vibrant than ever, more determined and ruthless—with himself as well as with others. He was obsessed with a need for haste. His throbbing impatience ruled out any possibility of careful consideration or involved calculation. He was becoming a man with no half-tones in his register of thought and action—a man who at times could be ludicrously dogmatic. He would recognize only good and bad, strong and weak, truth and untruth, and had no time for in betweens. A man whose mind was wondrously quick and clear, but not broad—a mind that could be at peace only with extremes and couldn't cope with nuances.

Myers the writer, the student of human character who vaguely sensed all this, had opportunity to confirm some of his theories later that night. After dinner the two of them went to the Orderly Room and the moment the door was closed Tuck dropped his assumed gaiety and unleashed the full blast of his urgent, cold-hearted disposition.

"Right, let's have it. I know the squadron's record—what a miserable shower of bloody deadbeats! Not worth a bag of nuts, far as I can see. Now—you're going to tell me which of them you reckon are worth salvaging, and which should be given the boot."

Jeff, loath to accept this responsibility, raised hands, shoulders and eyebrows in an ancient gesture, but before he could frame his protest Tuck smashed a fist down on the trestle table between them and positively snarled: "Look—if you're going to stay on here yourself, you'd better get on my side right this minute. I want the personal record of every pilot. I'll see their files later—first I want *your* opinions, the straight talk and no bloody nonsense. We'll start with the N.C.O.s." Jeff sighed, sank into a chair and began to talk in a low, tired tone.

Myers knew all right. He had a deep affection for every one of these young men and he understood their terrible disillusionment and bitterness. He had lived with them through their ordeal—all the confusions and nerve cracking defeats. He had watched their spirits ebb and their cynicism grow, and he had lain awake many a night because there was nothing he could do to help them. Even the recent replacements, some of whom had never seen an enemy machine, had been infected by the grim atmosphere.

He knew that the strength and sense of security of a fighter squadron was drawn from the pilots' familiarity with, and utter confidence in, one another's courage and capabilities. Without this, they were dead men. He had been forced to the sad conclusion that 257 was too far gone to be saved, that it could never be made to work as a team in that high, gay and trusting spirit which was so vital. But against his better judgment and sober nature, the exciting thought stole into his mind that maybe this was his chance to help them —maybe now, by talking with complete frankness to this man Tuck, he could save them. . . .

257 had been formed only a few months before. Most of the pilots had been drawn from other operational squadrons—men with plenty of flying experience, unquestionable ability and courage. But somehow they'd never come near to working as a team. The reason was simply that the squadron's previous commanders had failed to win respect and affection.

They had recently acquired the title "Burma Squadron," because the people of that distant land had volunteered to adopt a unit of Fighter Command and contribute money to pay for its equipment and upkeep. They'd been taking the money under false pretenses— in July and August they'd suffered brutal losses in men and machines and achieved a negligible "bag." A week or two ago they'd been "rested," and now they were back at what was by present standards reckoned as full strength—counting reserves, fourteen pilots and fourteen aircraft. But as a result of their calamitous engagements, the bickering and the mistrust, two of the sergeants were bad "twitch" cases, regularly reporting sick to avoid flying. They complained of stomach aches, back aches, stiff necks and other ailments which the M.O. couldn't dismiss because they didn't necessarily produce visible symptoms. One of these men had some excuse for his edginess: only a few weeks ago he'd been clobbered by a 110's cannon and got the backs of both legs peppered with shrapnel. The other was a replacement who'd yet to see a foe in his sights—simply scared stiff by the dismal state of the older chaps.

On the other hand, there was a third N.C.O., a thick-voiced Glaswegian named Sergeant Jock Girdwood, who was a keen and cheerful type and a very able pilot. One or two of the officers had "twitch" too —they said so, quite openly, and spent much of their time thinking up schemes for getting themselves posted to nonoperational jobs. It was an ugly picture, yet Myers was convinced that individually they were all basically tough and efficient. What they needed was what they'd never had: a true leader, who could impose discipline and inject faith. He named the most reliable of them—Pete Brothers, a flight commander; David Coke, son of the Earl of Leicester; the Canadian "Cocky" Cochrane, Jock Girdwood and a youngster named Farmer. There were four Poles, competent enough, but very independent—Franek Surma, Carol Pniak, Henry Sczensny and "Lazy" Lazoryk.

Tuck asked lots of questions about every one of them. He wanted to know not only each man's flying record, but his background, personality, hobbies, foibles and friendships. In the end he decided to give nobody the boot right away—they'd all get a chance to extract their fingers, but at the first sign of defiance, incompetence or Hun shyness he'd be utterly merciless.

It was after 2 a.m. when they left the Orderly Room to go to bed. Tuck carried with him a number of bulky files—the confidential records of the men he hoped to weld into an effective force. Alone in his room he read each one slowly, painstakingly, before he went to sleep.

Very early next morning he made his first flight in a Hurricane.

"My first reaction wasn't good. After the Spit, she was like a flying brick—a great, lumbering farmyard stallion compared with a dainty and gentle thorough-bred. The Spit was so much smaller, sleeker, smoother —and a bit faster too. It nearly broke my heart, be-cause things seemed tough enough without having to take on 109s in a heavy great kite like this.

"But after the first few minutes I began to realize the Hurri had virtues of her own. She was solid, ob-viously able to stand up to an awful lot of punishment . . . steady as a rock—a wonderful gun platform . . . just as well-powered as any other fighter in the world, with the same Merlin I knew and trusted so well. . . .

"The pilot's visibility was considerably better than in the Spit, because the nose sloped downwards more steeply from the cockpit to the spinner. This, of course, gave much better shooting conditions. The undercart was wider and, I think, stronger than the Spit's. This made landing a lot less tricky, particularly on rough ground. The controls were much heavier and it took a lot more muscle to haul her around the sky—and yet, you know, after that first hop, after I'd got the feel of her, I never seemed to notice this, or any of the other differences any more."

When he landed and taxied in, the ground crew chief, Flight Sergeant Tyrer, gave him a smart salute and reported: "All aircraft serviced, sir." He was a short, neat man in his forties with a chest thick as a keg, a ramrod back and hair cropped so short that the skin of his head showed through, deeply tanned. Long years of foreign service had furrowed his honest, sagacious face, faded the blue of his 1930-pattern uniform but colored it with a wad of campaign rib-bons. His buttons shone like sovereigns and his boots were smooth and glossy as ebony. He exuded smart-ness and efficiency. Tuck knew instantly he could count on him for first class service and complete loyal-ty.

Tyrer introduced members of the ground crew who would keep his Hurri (identification letter 'A') on top line. The instrument repairer, Leading Aircraftsman John Ryder, threw up a sizzling salute and blurted out: "The lads are very glad to have you here, sir." Tuck thanked him. Ryder seemed keen and capable.

He liked, too, the Leading Aircraftman fitter chosen to look after the C.O.'s engine—a shambling, grinning Cockney named Hillman who had a flattened nose, chipped teeth, a head shaggy as a hearthbroom and hands huge and ungainly as bunches of bananas. Hillman wore his forage cap with the side flaps dangling loose on either side of his face, and the droopy length of him was liberally splotched with grease. He greatly resembled a renegade spaniel emerging from a mudbath.

"Hillman's the scruffiest type of the station," Tyrer confided, "but I promise you, sir, he's also the best engine fitter." Tuck watched the L.A.C. go to work on the intestines of a shell-smashed Merlin, and knew this was true: those huge hands were strong as steel vices, yet the thick, scarred fingers were nimble as a woman's.

He was glad to see that the squadron had plenty of new fuel trucks—at long last the "turn around" problems were being tackled. Tyrer said he'd start competitions between the Flights to see which could get their aircraft refueled fastest.

Later that morning Tuck took the whole squadron off to begin an intensive program of battle drill which was to last, almost without pause, for three days. They snatched their meals out of "hot boxes" at the flight dispersals, grabbed a few minutes' rest on the grass during refueling.

"The first thing I had to do was to get them out of some of the old ideas which the other operational squadrons had dropped weeks, maybe months, before. For instance, they were still trying to fight in sections of three and four instead of in pairs, and their squadron formation was much too tight.

"After some general drills I started sending them up in pairs to fly a certain patrol line. Then I'd take off myself, lie in wait along the line and try to 'bounce' them from astern, usually out of the sun—Jerry's favorite trick. If I managed to get on to the tail of one I'd bawl over the r/t at him: 'You're getting it smack up your backside, chum! Your own bloody fault—why don't you take a look in your mirror now and then? C'mon now—try and shake me off!"

"Pete Brothers turned out a corker. He was highly intelligent and devoted to his job—an excellent flight commander. He picked up everything first time and helped me get it across to the others. Every now and again we'd take half the squadron apiece and have a ding-dong battle. It was then I found that as soon as they got excited they would overuse the r/t and forget all the set procedure—just as 92 had done back in the first May battles. The other great mistake was that they opened up miles out of range. Both of these were habits of long standing, not easy to break.

"During this period I never ordered them to do anything without explaining all the reasons. The more questions they fired at me after a flight, the better I liked it. I addressed them by their Christian names, or nicknames, but not too often—most of the time while we were in the air I didn't use names at all, just section numbers. If I'd tried to be too chummy, a lot of them would have shied away from me. As it was, I was pretty unpopular with a good many of them. On the other hand I couldn't use my rank to let off steam and shout some sense into them—I knew the big guns should be saved for the crises which must surely lie ahead.

"All the time we were drilling, of course, the pace of the big battle was mounting. Every day the Germans were putting up more and bigger formations —the summer was nearly over, and they were growing desperate. Fighter Command was strained almost to breaking point. I knew we hadn't much time.

"Now and again on these practice flights we'd see masses of vapor trails in the distance, high over the south coast or the Channel. It was a terrible temptation to swing round, belt over and join in. But for those few days the real war didn't concern us. We were still at school."

In the evenings: lectures in the briefing room, talk sessions in the mess. He showed them that most German bombers had their "blind spots." He had his fitters

build models, so that he could demonstrate to the
pilots approaches from various angles, and in every
sort of flying attitude. He made them work out the
different speeds, engine revs and throttle settings
needed for each type of attack. He fed them Intelligence
reports which analyzed the results obtained by other
squadrons, and handling notes on the latest modifications in machines and equipment.

At the start he'd thought it would take a week or
ten days to get them anywhere near ready, but towards the end of the second day's flying to his delight
all at once he found them clicking into position quickly, thinking much faster, keeping better look out and
—best of all—generally displaying a bit of dash and
initiative in the mock dogfights. And in the mess that
evening the majority of them seemed positively cheerful.

Myers was astounded by the metamorphosis. After
deep thought he attributed it to the simple fact that
Tuck so obviously knew his trade and fortunately had
the facility to explain things lucidly. The air force was
full of fine, hot workmen—but here was a rare one
who, in addition, was able to pass on to others his
knowledge, his enthusiasm, and even his skill!

On the third day the improvement was enormous.
Tuck signaled Group Headquarters that he was making good progress and should be ready for operations
in three or four days.

But they couldn't be spared any longer—the Battle
of Britain was reaching its climax and the reserve of
fighter pilots was pitifully low. Every trained man was
needed to oppose the Luftwaffe's final bid to smash
Fighter Command and to take full control of the
skies over the last European kingdom holding out
against the swastika. The next few days might well decide the issue.

An hour before dawn on September 15th Tuck was
awakened and handed orders which had just come
over the teleprinter from Group. He was to fly 257
to Debden, Essex, immediately. There two other squadrons would join him, to form a wing which he would

command. Good weather was forecast and heavy attacks were expected.

He was dressed in two minutes, and in less than fifteen Chief Tyrer had their engines running up. No time for breakfast—they swallowed black coffee on the tarmac and took off in semidarkness.

School was over.

★　　★　　★

The 15th now is history. This was the peak day, the turning point of the Battle—perhaps of the war. The Burma boys—the "dead-beats" and the "twitch" cases, called into action only as a last resort—did a very fair day's work. . . .

The bombers approached London from the southeast, flying at about 22,000 feet. As the wailing of the sirens rose to mingle with the malicious drone of the raiders' engines, people in Bromley and Dartford, Sevenoaks and Sidcup, looked up and saw ugly clouds of aircraft besmirching the immaculate September sky. The complex teleprinter networks of Fighter Command and Air Ministry and the operations rooms of a score of fighter airfields chattered out a top priority message which included the phrase: "Attack develop-

ing estimated strongest yet mounted in daylight." It seemed that the enemy had put every serviceable bomber he had into the air, resolving that this day should give Germany undisputed mastery of the skies.

To hold the whole raiding force close together the slowest aircraft had been put in the lead—a bunch of about 50 He. 111s. Behind, and a little to one side of these, followed very large formations of Dornier 17s and Dornier 215s. Finally, throttled back to a little above stalling speed in order to keep station, came a solid phalanx of about forty of the fast and formidably gunned Ju. 88s.

Between 5,000 and 8,000 feet above the bombers the escorting fighters fussed—a squadron of Me. 110s, and another of 109s. In all, there must have been over 250 "bandits" heading for the densely populated City area—lunging for the very heart of the British Commonwealth's commerce.

It was a hot, windless day. This had been a fabulous summer—the kind people had dreamed about before the war, but hardly ever seen. Ironic that now, just when rainclouds or gales would have provided respite for the desperately outnumbered defenders, day after day Britain should awaken to bright sunshine.

The raiders' plan was to drop their bombs all together in a comparatively small area, thus blowing a hole in the hub of the City. They reached the southern outskirts unchallenged, and turned for the run on to their target. It seemed that now nothing could stop them.

But, far below, the sun glinted on metal. Up out of the indigo haze that lay over the great capital climbed a wing of fighters—two squadrons of Spits, one of Hurris. Thirty-two little craft, rising resolutely into the vast blue battleground to face this Nazi Armada. And stenciled on the fuselage of the leading machine—a Hurricane—were fourteen swastikas, curt epitaphs for the German aircraft destroyed by its pilot—Bob Stanford Tuck.

Tuck's wing had already flown very hard that day. Three previous patrols, a total of over seven hours in the air, and all to no purpose—for while other wings plunged into the rivers of raiders, somehow Tuck's force had always just missed their interceptions. Constant effort and repeated frustration had drained their energy: now they were stiff, sweaty, eyesore.

But this time they could see their enemy. A lot of enemy, plenty for everyone. They jerked wide awake and aches went out of tensed bodies.

Everything was against them. There was no time to gain altitude or to get the sun behind them. If they were to break up this huge formation before it reached its target, then they must attack from underneath. This meant there could be no surprise, and that the German fighter escort would have the advantage of height which would allow them to dive on the defenders at considerably higher speed.

Tuck had ordered the wing to join up in a very loose formation of three "vics," but the planes which had been last to take off couldn't catch up and his whole force had become dangerously stretched out. Yet Tuck and the other leading Hurricanes dared not throttle back to wait for the laggards—to make this interception in time they must keep the levers slammed forward through the "emergency" gates and use everything in their engines.

Thus, when the leaders came up under the bombers and Tuck called his pilots into line abreast, only about seven Hurris were able to draw level with him for the all-important opening blast: his number-two man, Carl Capon; Pete Brothers, David Cook, Farmer, Lazoryk and Surma all very probably among them. It looked like a pitifully thin line. Eight of them formed very loosely, hurling themselves against such a host—tier upon tier of well-drilled aircraft, marching stolidly towards the teeming metropolis. But still Tuck didn't slacken off his battle-climb so that a few of the others could join them: every second gained would place the interception a few hundred yards farther away from the packed City, and increase their chance of saving hundreds—perhaps thousands—of civilian lives. . . .

The German fighters, of course, had seen them coming all the way. The 109s picked their moment, peeled off and came screaming down through the narrow gap between the close-packed formations of Junkers and Heinkels—which were still well out of the range of the Hurris. But Tuck and his seven companions kept right on, ignoring the plummeting 109s—not even firing at them as they flashed by directly ahead, but saving their ammo for the boys who carried the bombs. Somehow they survived that heavy spraying without serious damage.

And then they were through the fighter gauntlet and

closing on their prey. Tuck got an 88 lined up in his sights, but just as he was about to open up, a last and unexpected 109 came hurtling straight at him. Its stream of tracer stabbed very close over the canopy. He knew if he held to this course in a second he would die. There was nothing for it but to slam the stick hard over and climb away in a steep turn. Capon followed him perfectly. All at once, straight ahead he saw a pack of 110 fighter-bombers making a slow, wide turn —probably to face the rest of the wing, the two Spit squadrons, which by this time were drawing near the bomber stream. He picked one out, closed swiftly, checked the turn-and-bank needle and pressed the button. The 110 shuddered, tumbled out of position and burst into flames. At once the rest of this formation began to heave restlessly. Some of them skidded wildly, almost colliding with their neighbors. Others dropped their noses and dived blindly. Within seconds the whole pattern disintegrated. Tuck gave a satisfied grunt—this affair was going surprisingly well.

With Capon still in close attendance, Tuck swung round and half-rolled after a He. 111 which had suddenly panicked, broken away from its post in the stream and was fleeing for home. But again, as he lined up to fire, down on him—almost vertically, and incredibly fast—came a 109. Its bullets went wide, and it plunged past before he could reply.

"After him, Carl!" Capon dropped like a stone and Tuck, instead of trying to get back on to the fleeing Heinkel, craned his neck and peered up into the limitless, shimmering blue. Somewhere up there, he knew, a second 109 would be diving—the Germans, like the R.A.F., nowadays nearly always operated in pairs. After a moment he spotted the number-two following its leader down—but this second Messerschmitt pilot either saw the trap that had been laid for him or somehow became confused, for he pulled sharply out of his dive and crossed in front of the waiting Hurricane, flying into the center of Tuck's sight as obligingly as a homing pigeon to its loft. Turn-and-bank . . . a long burst. The 109 rolled smoothly on to its back and went down, smoking thickly.

Tuck went into a steep turn and looked around him, searching for another victim. But the battle was over. With astounding suddenness the sky was empty of all other aircraft. British and German machines had scattered, vanished, in what seemed an instant. After the whirling, twisting confusion there was a tremendous stillness, a palpable nothingness. All that was left were the tangled, filigreed vapor-trails . . . a few smears of oil smoke . . . a single parachute away to the south, floating tranquilly down into the motionless haze of the great Thames Valley. . . .

The Huns had been licked! That huge formation had been sent scuttling for home by a comparatively tiny force—a *magnificent* little force! He chuckled, remembering that the first assault on the main bomber stream had been made by only eight of them. The Burma boys had stuck together, and they had started the rot—alone, without the aid of the Spits. . . .

He was wonderfully proud. Proud of himself, but more proud of his squadron. The fleeing Germans would jettison their bombs at random and most of them would explode harmlessly in open country.

Back at base a jubilant Myers told him that 257 had been credited with the destruction of five machines. Without loss. The ground crew stenciled two more swastikas on their leader's aircraft, bringing his official total to sixteen.

That evening the 257 mess, which for so long had been a place of gloom and tensions, rang with exuberant cries, warm laughter, raucous singing—"Bang Away, Lulu," "O'Riley's Daughter," "The Ball o' Kirriemuir" and other charming choral works. Tuck was the pivot of the party, pacemaker in the drinking and the loudest—if not the most tuneful—of the songsters.

"That September 15th was one of the most important days of my life. It was the day that 257 became a squadron, and after that they never looked back."

11

On September 17th Hitler and his commanders met to decide the date for "Operation Sealion"—the invasion of the British Isles. Already the date had been postponed several times, and consequently the fleet of over 1,000 invasion barges waiting, crammed into the French Channel ports, had been receiving a nightly drubbing from R.A.F. bombers.

The trouble was that despite the Luftwaffe's claims that Fighter Command was "as good as dead," the Spitfires and the Hurricanes wouldn't stay down. Goering, the Luftwaffe's commander, and his deputy Erhard Milch, who were supposed to have delivered the "knock out blow" during the last few days, were forced to admit they had miscalculated—the exaggerated claims of their own air crews had misled them into believing that Fighter Command had been whittled down to less than a hundred machines!

In point of fact, by this time Britain's aircraft production had mounted spectacularly, under the brilliant leadership of the new Minister, Lord Beaverbrook, and was actually ahead of German fighter output. The R.A.F.'s problem was the grave shortage of experienced fighter pilots—the operational strength was down to around five hundred men.

Squadrons, on the average, now had only sixteen operational pilots out of their intended full wartime complement of twenty-six. Whole units had been wiped out. (There is a persistent legend of one squadron which, ordered to "scramble," signaled a smart acknowledgment and then sent up a single Hurricane, its sole survivor.) On September 8th Dowding, the C.-in-C. Fighter Command, had been compelled to introduce a pilot-economy plan known as "the Stabilization Scheme," by which certain squadrons were pulled out of first line service, reduced to five or six experienced men apiece and then used more or less as Operational Training Units to "bring on" swiftly the trickle of new pilots arriving from the flying schools.

It was fortunate that German Intelligence failed to discover this great weakness—had Hitler known, despite his own heavy losses he undoubtedly would have persevered for a few more weeks with his daylight bombing policy and very possibly destroyed Dowding's force. The decisions of September 17th are therefore of the highest importance.

On the conference table before the Nazi leaders lay detailed reports of the heavy and sustained air fighting of the 15th. On that day the Luftwaffe had mounted over 1,000 sorties: fifty-six German aircraft had been lost and only a resolute few had managed to squirm through to their objective—the Central London area. Goering's maximum effort had ended in defeat. They were forced to conclude that Fighter Command had adequate reserves of aircraft after all. It didn't occur to them that there might be no pilots to fly them. . . .

Admiral Raeder, commander of the German Navy, had no confidence in the ability of the Luftwaffe to give his invasion fleet adequate cover on the way across to the English beaches. He complained that even now, with his vessels moored in well defended harbors, he was getting very little air protection—indeed, the British bombers had recently sunk over eighty barges in one night. Understandably, the Admiral suggested: "The present air situation does not provide the conditions for carrying out the operation, as the risk is still too great. A decision should be left over until October."

There was a long and no doubt heated debate, full details of which may never be known. But the outcome is tersely recorded in the official German War Diary: "The enemy air force is still by no means defeated; on the contrary it shows increasing activity. . . .

The Fuhrer therefore decides to postpone 'Sealion' *indefinitely*."

Thus ended the Nazi bid for air supremacy. It must have taken a lot to convince Hitler that his beloved Luftwaffe had failed him—that mass daylight attacks hadn't worked, and therefore they must now resort largely to night bombing. He'd always supported Goering against the other commanders, but there was no denying the frosty facts of that report on the conference table, headed: "Air Situation, September 15th."

* * *

The last day assaults by big formations of bombers came in mid September, then the tactics changed markedly. An occasional lone Ju. 88 or He. 111 struck boldly in the full light of the autumn sun, but now they came *en masse* only by night.

Between dawn and sundown the attacks were almost exclusively fighter raids—huge packs of Messerschmitts 109 and 110, with a few of the new Heinkel 113s. For hours on end they'd keep coming, in waves of twenty or thirty. Most of them were at between 25,000 and 30,000 feet, but some carrying small bombs on improvised racks under their wings arrived with the Channel's salt on their tails, having darted over at sea level in the hope of avoiding radar detection.

The frequent, hectic duels with the fighters and fighter-bombers meant even longer working hours for the defenders, but pilot losses were slightly less than in the recent encounters with the bombers and mixed formations. Thus, gradually, Dowding's "stabilization scheme" began to show results: operational pilot strength ceased to fall, as more and more youngsters arrived to fill the empty cane chairs outside the flight dispersal huts.

This—the final phase of the Battle of Britain—was the period of greatest strain. September blazed like June, and autumn would not come to their relief. . . .

* * *

You were awake long before daybreak . . . busy in your cockpit as the first chill slivers of light thrust up out of the enemy's domain to the east, and usually in the sky while the sun still clung sleepily to the rim of the earth.

Scramble—climb—vector—*buster!* Your life was packed with action, the breathless, throbbing sensation of intense danger. Save when the weather was foul, you flew and fought the deathlong day.

Very occasionally, perhaps, in the midst of the flurry came an hour or so of deep, waiting silence, with the autumn breeze blowing across the airfield, bringing the scent of roses and honeysuckle to mingle with the tang of oil, rubber, petrol and the fuselages frying the sun. Then you lay down on the grass and slept: tousled heads on parachute packs, everyone curled up like an Arab in a sand hollow. . . .

Not letting yourself think too much about the newcomers, the young boys fresh from the flying schools, calm and flippant at first, and then pathetically surprised to find themselves killing and being killed. Not letting yourself think of home too often, not writing the family about any of this. Just abandoning yourself to the raging routine, and—above all—being fiercely cheerful all the time, to prove you weren't afraid.

You dragged yourself into the air more exhausted each morning, and you came back at nightfall too keyed up to rest. Long after you landed there was a roaring in your ears, so you shouted and drank, and you laughed and sang and drank, with your pals in the mess or at the local until you all got a bit drunk. At last you got back to your little room and crawled between the soft, white sheets tired, tired—your head, your muscles and your bones aching. And you found you couldn't sleep because of the sudden silence, so you just lay there alone in the darkness and it was then that your nerves began to tauten and twitch inside their casing of beaten, numbed flesh. It was then that you were forced to listen to your thoughts, to take a tally. . . .

Every day now, somebody you knew got the chop— on another squadron, yes maybe, but usually somebody you'd always considered a better pilot than you.

Secretly, so secretly that you kept it from yourself most of the time, you'd faced the fact that your turn was bound to come soon. There was a noose around your neck—it was only a question of time, you were just a number that would be wiped from the slate tomorrow. Or the day after tomorrow. . . .

At the end of each day's ordeal, the Battle of Britain pilots returned to the quiet and the comfort of the English countryside—went into the old world pubs and rubbed shoulders with farmers and shopkeepers and ditch diggers and solicitors whose way of life had scarcely been touched by the war. Or perhaps they sat in the cinema watching Betty Grable or Bob Hope. And afterwards they lay down to sleep on sprung mattresses between soft sheets. They were many miles from the nearest enemy troops, occupying comfortable quarters in the midst of the civilian populace, within earshot of village clocks which had chimed through safe, humdrum centuries. In the evenings they were part of a fairly normal, calm and conventional community. And yet by day they lived in a remote, savage world of their own, fighting a ceaseless battle of nerves and wits, and every morning the transition grew harder.

Such circumstances didn't exactly encourage the virtues of decorum and self-control. When the day had been survived they hurled themselves into their pleasures with a frantic, animal zest and the same recklessness that served them so well in their work. And since those who ply the gladiator's sorry old trade ever acquire a precarious sort of glamor, the hotel bars around the fighter bases were frequented by droves of young women with wide, inviting eyes—girls with quick, hard laughs that hid either maternal pity, or that strange, almost masochistic anticipation—part dread, part thrill—which through the ages has prompted so many women to give themselves to condemned men.

The girls found these boys eager, determined, but usually artless. Quite often, they would get too drunk to make love well. They liked to stay in a crowd, making a lot of noise, until the bar closed. They hated to be alone for long, even with a girl, because that meant going somewhere quiet. All the day they hurtled about the sky with the commingled dins of battle beating at their heads—thus quietness had become their enemy, tempting them to slacken pace, to stop and think—confusing them, undermining their strength. They could not be alone, they could not live for more than a few minutes with silence. They were hard at war, so they could not be at peace with peace.

Even in their cups they were never introspective—they couldn't afford that. They clung to each other —gloriously, if deliberately, extrovert. They had a tremendous understanding of one another, and astounding vitality. Their foamy, spendthrift gaiety seemed to do them little harm, for they drank practically nothing but beer, and the slipstream cuffing their heads as they started up their engines in the chill of early morning swiftly cleared the cobwebs.

Every fighter squadron was like a tribe, with its own chieftain, customs, songs, rituals and crest. But their off-duty interests never varied. After the 15th, the pilots of 257 seemed to throw themselves into this way of life—which was, by Fighter Command standards, of the time, "normal"—and, headed by the tireless, swashbuckling Tuck, they shook the tiles and timbers of every hostelry within ten miles of Martlesham.

To keep his place, to hold them together in the surge of this great battle, Tuck had decided he must fly more than any of them, take greater risks, laugh more and drink more. Charcoal smudges under his eyes were the only clue to his exhausted condition. He kept telling himself that very soon the pace must slacken, he would sleep later, sleep for the rest of his life—but not now, not when every hour vibrated with drama, challenge, opportunity. It didn't cross his mind that the unruly force which drove him on might be nothing less than the fear of losing his reputation, his self-confidence and skill—that same old Grantham ghost, still snapping at his heels.

No man ever worked harder at his trade. And it seemed he was at his best when stretched to these limits of endurance.

And yet, for all his success in his first days with 257, Tuck was far from satisfied. He could sense that some of the pilots were hanging back, withholding the full strength of their personalities from the team, both on the ground and in the air. These included the two N.C.O's he'd listed as "twitch" cases. To encourage them, he'd had both switched to his own flying section and spent a lot of time talking with them. Yet they remained withdrawn, stiff, formal. He thought it might only be that they felt guilty about the past, and that in time they'd loosen up.

One morning the squadron was vectored on to a mob of about forty high-altitude 109s at 24,000 feet over the Thames Estuary. (At this height the Hurricane lost the edge of her performance, becoming distinctly sloppy on elevators and ailerons, apt to "mush" and wallow if the pilot moved the stick too violently, and generally slower in speed and response. Normally the Spits worked the "high beats," but nowadays the planners' best laid schemes often went awry, and the Hurris had to fight uneven battles in rarified conditions.)

They had been joined by 73 Squadron, and Tuck was leading. The 109s made a sharp turn to the north and dived hard, along the coast. All to the good—the lower they went, the better. But in following round, the Hurri wing got strung out, so that the first section to get within range of the Messerschmitts was Tuck's— Carl Capon and the two sergeants. As the altitude decreased the Hurricanes began to gain slowly.

"Line abreast—good and wide!"

But as they drew within maximum range, out of the corner of his eye Tuck saw one of the outside machines abruptly rise on to a wingtip and peel off. An instant later the other outside man followed.

The sergeants had funked it!

The shock of it made him reel in his harness. He moved his lips but the only sound he made was a hoarse grunting. He realized there was no point in trying to call them back, because already they'd be two or three thousand feet below and too far behind to catch up even if they obeyed his order. So he just sat there, grunting, icy cold all over, the flesh of his face paralyzed in hard knots.

He mustn't go tense like this—this fury could befuddle as thoroughly as panic. He had to forget the bastards and concentrate, concentrate. . . . It was going to be very dicey now, with just the two of them going in and the others strung out well behind—if he and Capon passed into the middle of that pack and got cut off from the rest of the wing. . . .

"Carl. . . ." He found his voice—a shade softer and higher than usual. "Carl, drop back a bit and cover me. Quickly, now."

The other Hurri slid smoothly into position behind and slightly above on the port side. They closed to eight hundred yards, seven hundred, six-fifty. A three-second burst. A 109's wings rocking, ever so gently, as the chains of flaming metal transfix it. The slow drop of the nose, the long dark streams coming back, like entrails, as it starts down on its last, wailing dive.

Tuck turned quickly after a second victim, but Capon's voice suddenly jumped an octave: "Behind you! On your right!" A single 109, lingering behind its fellows, had sliced in brilliantly from above and fastened on to the Hurricane leader. Even as Tuck started to pull up sharply and turn tightly to the right, hauling with both hands on the stick, the first blast struck him.

Harder round, you fool! Tighter, faster! Want to live? Sweat blood, then! Fly, fly!

A second burst hit him: a stunning clatter, violent shuddering and vicious flashes all around him. The glass reflector plate of the sight disintegrated a few inches from his face and the wind shrieked through jagged holes in the canopy. But still he held her in that tight turn, and suddenly he was out of the welter of bullets, and his engine was still roaring staunchly. He flattened out quickly, but in those few seconds his attacker had vanished and the enemy pack had broken up. There were odd pairs and threes and fours streaking off in all directions, some climbing, a few turning steeply away to the north or south, but most of them were diving out to sea, with the Hurricanes squirting at their tails. It was futile to try to catch up with the running fight.

"Home." *And gentle Christ in Heaven help those funking sergeants.*

They were waiting on the tarmac, standing together, chalk faced and fidgety. Tuck got out of his damaged aircraft and walked very slowly towards them, watched by two or three other pilots and nearly all of the ground crews. There was a white, smoldering rage in the depth of his being. His eyes were vicious black pebbles. Several feet away from them he halted, breathing very slowly and deeply, his sun reddened nostrils quivering.

And then suddenly he stooped and straightened in one smooth coiling movement, and came up with gleaming metal in his right hand—the Mauser pistol, salvaged from a crashed Heinkel weeks before, which he now carried in the leg of his right flying boot.

He stepped quickly closer, the pistol leveled at them. His face seemed thinner than normal and the scar on his right cheek stood out like a welt of dead, white flesh. He began to speak, quite quietly but with assiduous pronunciation, spacing his words very evenly.

"In time of war," he said, "desertion is as bad as murder. Sometimes it *is* murder. It bloody nearly was today. You two funked off and left the rest of your section." The hand holding the Mauser was raised an inch or two. The sergeants went rigid. "You deserve to be bloody well done! A bullet apiece—that's all you're worth!"

Chief Tyrer moved swiftly to Tuck's side and reached for the gun. But he didn't touch the weapon, merely held a hand above it, palm down.

"Easy, sir." His voice was relaxed and respectful—exactly the same tone he'd have used to report "All serviced, sir." After a moment Tuck lowered his arm and took a long, deep breath. Then he jammed the Mauser in his pocket, walked up to the sergeant who'd been first to peel off and snapped: "What've you to say for yourself?" He was a gangly, droopy youngster with a blotchy skin, and he was trembling so much that he could hardly control his lips and tongue. In a jerky, breathy babble he declared that he'd suddenly noticed a serious drop in oil pressure—just as they were going in to attack. . . . It was so ridiculous that he broke off in the middle of a sentence, let his head fall and stared at the ground, beaten.

"Flight Sergeant Tyrer, place this man under close arrest." While an escort was being organized to take the prisoner to the guardroom, Tuck turned to the other N.C.O., a youngster of nineteen with wavy, buttermilk hair and a very handsome face. This was the boy who'd suffered leg wounds before Tuck took over the squadron. Misery swam in his eyes, but he stood to attention and his voice was steady enough.

"No excuse, sir."

"You deliberately broke off?"

"Yes, sir."

"Why?"

"Because I got 'twitch,' sir."

"Before we'd fired a shot?"

"Sorry, sir. There were so many of them, I just couldn't make myself go on."

Somehow Tuck knew this boy had struggled with his fears, but failed. Probably it was seeing the other sergeant break off that finally defeated him.

"Go to your room, and stay there. You're under open arrest. I'll talk to you later."

"Yes, sir."

The sergeant saluted and left, walking quickly, his

unwieldy flying boots thumping on the concrete tarmac in the awed silence. None of the others stirred while Tuck lit a cigarette and took several pulls at it, staring out over the field. Then suddenly he spun round in a half-circle and hurried to the flight hut. The pilots followed and the ground crews drifted back to their posts, murmuring secretively.

The sergeant who'd been first to peel off was eventually court martialed and reduced to the rank of A.C.2. Many weeks later Tuck saw him at another airdrome where 257 were rendezvousing with other squadrons to form a wing. He still wore his pilot's brevet, but no other insignia. He was in the control tower, sweeping the stairs.

Tuck talked to the other youngster and, because he was honest and because he'd had a bad time at the very start of his operational career, decided to give him another chance. It paid off: eager to wipe out his shame, the boy overcame his nerves and very soon developed into a competent and reliable pilot. In time he was granted a commission.

The incident was the last of its kind on 257. The one or two "suspect" officers quickly pulled themselves together. The squadron spirit rose to new, boisterous heights, and it seemed that now they were a complete, unconquerable team.

Then came the trouble with the Poles—trouble of a very different kind. Now that 257 was operating so well, the four Polish pilots became so bloodthirsty, so fanatical in their lust to kill Germans, that once or twice they failed to obey their leader's commands and went off on tail chasing sprees of their own. On one of these occasions they followed a group of Messerchmitts so far out over the North Sea that they almost ran out of fuel and barely managed to get back and force land at a tiny, coastal training field.

Tuck liked the Poles. From the start he'd got on well with them, because to their great surprise his knowledge of Russian enabled him to understand almost everything they said to each other. The languages are no more dissimilar than, say, the sort of English spoken by Londoners and the version used in

the Scottish Highlands—the swear words are more or
less the same! When the Poles got excited in the air and
lapsed into their native tongue, their leader would
break in and curse them roundly in the accent of a
Ukrainian ditchdigger, but after a while even that
couldn't hold them in check—they became hurtling
bolts of fury, beyond all reason and authority.

One evening in the mess he turned abruptly to one
of them and said, not unpleasantly, "Oh by the way,
you and your balmy chums went tearing off on a
private war again today. I'm getting thoroughly
browned off with these capers. You started it this time
—I heard you. So I'm grounding you for twenty-four
hours. Tomorrow you don't fly."

The Pole was aghast. He apologized, he pleaded,
but Tuck was adamant. Hopeful till the last, the Pole
turned up at dispersal in the morning, wearing his flying gear. Tuck ignored him. When the squadron took
off at dawn, he was left standing on the tarmac. According to Jeff Myers and Chief Tyrer, big tears were
rolling down his cheeks.

After two or three doses of this medicine, the "Private Polish Air Force" became malleable, and was
duly welded into the team.

$$\star \qquad \star \qquad \star$$

In the deadly game of aerial warfare, Fortune seems
ever to favor the boldest players—and Tuck most of
all. Now, only two weeks after the death of Caesar
Hull, there ambled into the 257 mess an easygoing,
plumpish, slow-drawling Canadian who was to become an even closer and dearer friend. This was a
time when Tuck could easily have got careless, or even
deliberately exposed himself to ridiculous dangers—at
long last the pace was beginning to tell and exhaustion was confusing his busy brain. There was no
one now in whom he could confide, no one he trusted
and liked enough—no one who could take his mind
off his troubles and responsibilities for even an hour
or two. There could have been no more perfect timing
for the arrival of Flight Lieutenant Peter (Cowboy)
Blatchford.

Cowboy was posted to the squadron in answer to a request from Tuck for a thoroughly experienced flight commander. He had come over from Alberta and joined the air force a year or two before the outbreak of war, and had seen a good bit of action on Hurris. He was cheery-faced, chunky and chuckle-voiced, with an extraordinarily large backside that made him waddle and roll like an overfed puppy. The slowness of his movements and mannerisms proved wholly deceptive—his mind was rapier-swift, his reflexes instantaneous. He was a brilliant shot, never got excited—all told, a "natural."

From the start they'd only to look at one another and they were overtaken by a tremendous sense of fun. Even their grimmest and most urgent tasks were planned and executed in a spirit of banter and ribaldry. Cowboy was the only member of the squadron who, in the middle of a battle, could tell the owner of that flat, curt voice to "get stuffed" and get away with it—maybe even earn a long, loud and friendly raspberry in reply.

In presence of the other pilots he addressed Tuck as "boss," hardly ever said "sir." In private it soon became plain Bobbie, or Tommy. Sometimes it was "Beaky"—a dig at Tuck's long, thin nose. Tuck called him Cowboy most of the time, other times it was Pete or occasionally "Fat Arse." But though they never ceased to tease one another, somehow Blatchford always showed sincere respect for his leader. It was impossible to be with the pair of them for more than a few minutes and fail to see the affection they shared.

Apart from flying and beer, they had practically nothing in common. Cowboy had two absorbing interests: hillbilly music (which probably helped to earn his nickname) and girls. Yet it could be said that to a remarkable extent the Canadian "humanized" Tuck—got him to listen to, and even join in with, a host of twangy, corny records of "mountain music" and traditional songs of the Wild West, and once or twice even persuaded him to make up a foursome in Ipswich or London with "my girl's pal—real cute she is, honest. . . ."

The hillbilly sessions and the blind dates jerked him out of the work and worry rut. Though perhaps not wholly enjoyable, they were new experiences for him, and therefore they required all his concentration, if only for an hour or so. He returned to his work refreshed by change.

October laid frost on the hedgerows and rooftops, the days grew shorter, and now and then a low overcast, prolonged rain or mist bogged down the battle for a few hours. There were delicious mornings when they were able to lie in bed, or lounge around in the flight huts warming their hands on coffee mugs.

"Tuck's Luck" and a piece of brilliant ground controlling combined to destroy a lone raider on the 4th. It was a filthy day with solid, kinky gray clouds drawn low over the airfield like a rumpled and dirty tablecloth. Flying had been "scrubbed" but Tuck happened to be hanging around the big bell tent that served as dispersal H.Q. Suddenly a call came from the area ops room at Debden, a nearby airfield:

"We have a clear plot—bandit heading for your airfield! He ought to pass smack across you in just about two minutes. How's the weather over your way?"

"Perfectly bloody. Ten-tenths, right down on the deck."

"Same here. Any chance of getting something off?"

Tuck ducked his head to glance out of the tent flaps. Everything dismal, dull and dangerous out there. No ground crew in sight—and all the other pilots were in their rooms or at the mess. Control burst in again while he was still thinking.

"It's a *wonderfully* clear plot! If you could just get something up we could vector bang-on to him! This bastard's made a circular tour of England—if we've missed him once we've missed him ten times. You're our last bloody hope. . . ."

"Righto—let's have a bash!"

He slammed the phone down, snatched up his gear and sprinted to his aircraft. Precious seconds were wasted tearing the canvas cover from the hood and starting it up himself, but the moment the engine had fired, L.A.C.'s Hillman and Ryder came racing along the perimeter track on their bicycles. Hillman dived under the wing and whipped the chocks away. Ryder's fingers fairly flew as he grabbed the straps of Tuck's harness, yanked them tight, inserted the clip. No time for a proper check—every second counted. Fortunately all the engines on the line had been warmed up earlier. Before he'd even got going on his take-off run, he was tuned in to the controller's urgent directions. Now the plot was showing the Hun as almost over the airfield!

The Hurri sloshed through big pools of water and lurched into the air. Almost immediately he was in dense cloud, flying by his instruments. Control was shouting a northerly bearing, so he made a quick turn to port. He still had full climbing power on when the controller burst in again: "It's coming right on! The plots are close! Keep your eyes peeled." Tuck was still fumbling with his oxygen mask, trying to clip it on to his helmet so that he might use the microphone, when he popped out of the murk into dazzling sunlight. Directly ahead, a couple of hundred feet higher, a Ju. 88 floated serenely on an easterly course.

He threw the Hurri into a vertical bank and yanked her round to the right. This brought him directly under the raider. He throttled back a bit and stayed there while he switched on his gun-sight and flicked the safety-ring on the gun-button to "fire." Then he opened the throttle wide again and clambered up towards the 88's belly.

But as he drew within range the German suddenly stuck his nose down and dived—probably he'd spotted the Hurri's shadow converging with his own to the blotter smooth cloudtop beneath. It wasn't a steep dive, but Tuck knew that if he failed to make his kill before the raider found cover, there would be no second chance.

He climbed frantically, hanging on his prop, and put a long burst into that smooth belly from only a hundred and fifty yards. It was as if an explosion occurred inside: black smoke came billowing out of the fuselage and whole sections of the bodywork seemed to peel off and come whirling backwards and downwards, passing perilously close to the Hurricane.

Tuck fell back a little, and the rear-gunner gave one very short burst in reply. The tracer snapped by a few inches over the fighter's port wing. Then there was another big belch of smoke and the 88 shuddered, shocked from tail to nose, and keeled over. Tuck broke off, making no attempt to follow it down—it would be too easy to collide with it in that thick fleece below.

He called up control, and for once his voice on the r/t sounded excited.

"Pretty sure I got him! He's on his way down, bags of smoke and filth. . . ."

"Bloody marvelous show!"

"Bloody marvelous controlling!"

He flew on to the east for a minute or so, until he knew he must be over the sea. Then he let down very carefully. At around four hundred feet he came out into the gray drizzle and, turning back west, soon crossed the coast and picked up a town which he instantly recognized as Southwold, Suffolk. He was approximately six miles from base! Just off the beach there was a big foaming patch in the becalmed, pewter-colored water. He knew the Junkers lay in the darkness beneath.

The whole action—from the moment the dispersal phone had pealed, to the instant when the German exploded into the sea—couldn't have lasted more than five minutes. Tuck's logbook entry for this combat reads: "Intercepted and shot down one Ju. 88 off Southwold. Extremely good controlling." For months he had cursed all ground controllers as a snail-witted and treacherous race. He now realized that in the last few weeks they had made astounding progress. On the 15th, he recalled, they'd been extremely competent in handling large-scale attacks which had developed in very rapid succession, sometimes simultaneously. After today's wonderful piece of cooperation, he found himself thinking fondly of those gray-headed men who ran the ops. rooms—the victors of old air battles, now limited to playing a grim and intricate game of chess on the big plotting tables—with the sky for their board. A game in which, more often than not, the pawns made the decisions and became the true players, and the plotters were helpless spectators.

A few days later he was awarded a bar to his D.F.C. He was very surprised. He really meant what he said— "I've just been bloody lucky, that's all."

*   *   *

The squadron moved to North Weald, Essex, about the middle of the month. At this time, on a sudden whim he gave himself a day off and flew to Biggin Hill, where 92 was now stationed. He got a wild welcome from his old comrades—but they were tragically few now, and the mess was full of strange young faces.

The old gang were reminiscing at dispersal when suddenly an alarm came through. Tuck grabbed Kingcome—still a flight lieutenant, but in fact the air leader—and asked: "All right if I come along?"

"Pleasure. But not in that bloody Hurri. Take one of our spares."

They got off very smartly and control sent them climbing out to sea. The sky was clear and quite empty. Soon Tuck, excited by the feel of the sensitive Spitfire after nearly a month of Hurrilugging, fell to thinking of his first days with 92 . . . of Bushell and Learmond and John Gillies and the rest.

He looked out at the echelon and thought with vague, unreasonable anger: "Too many bloody strangers. . . ." What he really meant was: "What are those strangers doing in my friends' places?" He was hostile because their presence brought home the absence of the others, the comrades of—could it really be only three or four months ago?

Several had gone only in the last four weeks since he'd left the squadron. Johnnie Bryson . . . Eddie Edwards . . . Peter Eyles . . . Bill Williams. . . .

And Howard Hill—his had been a strange ending. He'd been hit in a scrap just off the French coast, and his damaged machine was seen to break off and turn for home, losing height. Pilots of other squadrons saw him struggling over the coast, but all attempts to contact him by r/t had failed—his radio must have been knocked out.

The crippled Spit reached Biggin Hill, but instead of coming in to land it passed a couple of miles to the south at about 1,200 feet, maintaining a steady course inland and gradually losing height. The Biggin Hill controller tried to call him up, but again there was silence. The fighter flew on to the west—and vanished. For a month they found no trace of it, then the pilot of an Anson trainer spotted wreckage in a dense wood only seven miles from the airfield. A recovery team went out and found Hill's Spitfire, lodged in the treetops over forty feet up.

They got ladders and climbed up. As they drew near the wreck they knew that Hill was still in his cockpit—there was a sweet, sickly smell that almost overpowered them. The body had been exposed to the sweltering sun for nearly thirty days, and under the thick perspex the temperature must have soared to hothouse level.

When they reached the top and looked into the cockpit, at first they thought the sun must have burned him literally to a cinder. Beneath the canopy they distinguished only a black something—and then, on closer inspection, they realized with horror that the body was sheathed in a crawling mass. Countless flies and other insects of a thousand species had entered through what was obviously a shellhole in the side of the hood.

Eventually it was established that a cannon shell had come through high on the port side and neatly taken off the New Zealander's cranium, like the top of a boiled egg. Howard, of course, had died instantly, cleanly—but his Spitfire had lingered . . . turned west, as if trying to reach home, somehow found its way back to the Biggin Hill area. It must have made a very good landing on the forest's roof, for the damage proved comparatively light.

On this same sad theme Tuck looked back still further, to Grantham. Caesar . . . Dougie Douglas . . . Eddie Hollings . . . Buti Sheahan. . . . All had folded their wings. And from Hornchurch—the *old* Hornchurch that had been shared by 65 and 74? Desmond

Cooke . . . Chad Giddings . . . Johnnie Welford . . .
George Proudman . . . Jack Kennedy. . . .

It was a long list, and getting longer every day.
Were *all* the regular officers to go—would none of the
prewar gang survive?

In a deliberate effort to shake off this gloom, he
started to count up those regular, or auxiliary, pilots
who, like himself, were still going strong: Sailor Malan,
piling up a high score and proving himself a born air
leader; the amazing Doug Bader, fearless and bel-
ligerently righteous, now leading huge wings and grinding out violent memoranda to his superiors on tactics
and supply matters; the lanky, laughing Canadian,
Johnnie Kent, volunteering to fly into steel barrage-
balloon cables in order to test experimental cutting
devices fitted to his wings; the grinning "Killer" Mac-
Kellar, who'd shot down five in one day; handsome,
hard-driving Max Aitken, son of Lord Beaverbrook
and one-time airline pilot in the United States, still
setting a cracking pace for the lusty lads of 601
Squadron; the rough, tough Al Deere, ex-boxing champion who'd K.O.'d over ten Germans.

And he mustn't forget the first Spitfire pilot ever to
ditch successfully in the open sea, Norman "Green-to-
Black" Ryder—so called because he included in his
description of his "underwater flying" adventure the
phrase: "as the kite went down, at first everything outside the canopy was a very pretty green color, then
it turned to black and I began to get worried." Then
there was John "Cat's Eyes" Cunningham, baby faced,
rosy cheeked and blond, droning through the foulest of
nights with a kiteful of mysterious equipment while
the other chaps were snug by their stoves, developing
a whole new technique for fighting the bombers in
darkness . . . Paddy Treacey, Johnnie Loudon, Mike
Lister Robinson, Peter Hillwood, all still merrily at
work.

Ah, they were still a power in the land—the "Old
Guard;" they still had a long way to go and many wonders to perform!

He was jerked out of his reverie by a black dot
which entered the field of his left eye, moving swiftly

far below. He leaned forward and watched it for several seconds as it grew larger and more distinct. Me. 109.

"Tally-ho!"

He completely forgot that he wasn't the leader here. He dropped out of the formation, dived at full throttle and wheeled nimbly on to the lone fighter's tail. The German couldn't have seen him, for he made no effort to escape. Two short bursts dismembered the tailplanes. The 109 switchbacked and zig-zagged like a drunken bluebottle, then went spinning down with long flames licking out of it.

Kingcome's voice filled Tuck's helmet.

"Well, stap me! Of all the bloody cheek. . . ."

Back at the Biggin Hill mess, over their bitters they argued on whether the 109 should be credited to Tuck or added to 92's total. Brian suggested that they halve the kill, but in the end: "Damn you Robert—take it! But I swear if you come back here you won't fly with us again. Bloody embarrassing—having to report a 'kill' to Intelligence, then explain that it was made by a visitor!"

*　　*　　*

Later in the month his sister Peggy arrived unexpectedly at North Weald for a weekend. She looked thinner, paler, but quite self-possessed. At first there was some uneasiness between them, but on their first evening out together—in the cocktail bar of a nearby hotel—she suddenly revealed that she knew precisely how he had been involved in the death of her husband. She didn't reveal *how* she'd learned it—in fact, he still doesn't know who told her."

"I've known for a long time now," she said. "Don't worry, I'm not an idiot, Robert—it would be perfectly ludicrous to blame you. I only mention it because —well, because I've been afraid you might let it prey on your conscience, or something silly like that.

A minute later, as he took the empty glasses to the bar, and was conscious of a kind of rising gaiety. The relief was quite heady. For the first time in his life he winked at a barmaid and called her "Sweetheart."

12

The night was cold, and very still.

As he flew through the dregs of the dusk from North Weald across London towards Northolt airdrome—where he was to attend a routine conference very early next morning—far away on the southern outskirts he saw a solitary searchlight probing, like a great, gleaming glass rod. The searchlight set him thinking about the grim winter to come—long, black nights filled with the thrum of raiders' engines, the teeth-rattling crunches of the bombs, the ravenous hissing and crackling of incendiaries taking hold of rooftops and walls. . . . Month upon month of destruction and death and clogging débris and growing shortages . . . of women and children having to sleep in bare, dank shelters or packed like animals in the tube stations. . . . These miseries would surely come.

Germany's summer assault had been beaten back, the R.A.F. had won the daylight battle; but hunting bombers in wintry night skies was an infinitely more difficult matter. Even with the new, rapidly organized and specially trained night-fighter squadrons—Defiants and Blenheims mostly, and a few of the new Beaufighters, all stuffed with the mysterious sort of equipment that John Cunningham had been testing for so long—the chances of stopping more than a fraction of the bombers getting through to their objectives after dark seemed to him exceedingly remote.

He banked the Hurricane and peered down at the dark mass of the city. Greater London, the largest target on Earth; 720 square miles in area and no part of it more than one hour's flying time from the new bases of the world's biggest air force. In the coming nights how many of her citizens would die? How many of her venerable buildings would become fierce pyres—how many of her ancient halls and churches would glow like lanterns?

On landing at Northolt he decided not to go to the mess: this station was frequently visited by older, high ranking officers from Air Ministry and II Group Headquarters at Uxbridge, nearby—the atmosphere usually was a bit too stiff and formal for his tastes. He was free until the conference in the morning, so he borrowed a car and set out for the West End. He would go to *Shepherd's*, the Mayfair pub which was the acknowledged rendezvous of fighter pilots on leave in London. There he'd be sure to find a bunch of the boys—there he'd soon snap out of it.

As he checked out at the gate, the service policeman on duty told him that the air-raid sirens had wailed their warning just a few minutes before he'd touched down. Driving through blacked out, near deserted streets he couldn't help glancing up occasionally at the night sky, but he saw nothing save a few more probing searchlights and a high, insolent moon—a cold fleck of a moon that seemed already to hold the glitter of frost. He heard no bombs fall, but several miles away, somewhere south of the river he guessed, a heavy A.A. battery began firing, and the air of the city was filled with soft jolts.

As he drew near the West End the streets suddenly became busier. Yet, the scene was colorless and shabby. The war's crudeness had swept away nearly all of the light and color.

By the time he reached *Shepherd's* he felt worn out, both physically and mentally, but the moment he'd fought his way through the black-out curtains into the warm haze and babble, a voice roared: "Hiya, Tommy —over here, you blighter, and buy us all a drink!" It was Tiny Vass, and with him round the far corner of the bar were Johnnie Kent, Max Aitken, young Peter Townsend, "Cocky" Dundas and several others, all broad grins and friendly eyes and clinking pintpots.

There were so many slant-striped D.F.C. ribbons that the room seemed tilted to one side. His gloom evaporated as he started towards them, squeezing through the densely packed throng of drinkers. Then, just as he reached them, above the din of the place came a lowrumble, and the windows rattled in their frames —a bomb exploding in the distance. But nobody took the slightest notice.

*   *   *

The big formations of fighters and fighter-bombers continued to come over by day during the final months of the year, and 257 were kept busy. Cowboy and Peter Brothers were positively brillant flight commanders, and the only worry was that Group's "talent scouts" would "discover" them and whip them away to command squadrons of their own! The "Burma" boys worked in happy partnership with 249 Squadron, commanded by John Grandy. A third, less experienced Hurri squadron made up the wing.

The station commander was Group Captain Victor Beamish, an officer with many years service to his credit and a rich, Irish burr to his voice. He was tough, direct, demanding. Though middle-aged, he was still flying, and insisted upon taking part in some of the most difficult actions assigned to the wing.

Things seemed to go very well for a while, and Tuck added to his personal "bag."

He was credited with one "kill" and two "probables" in a huge, churning dogfight over London in which the wing routed a swarm of about fifty 109 fighter-bombers.

On another big day the wing intercepted and scattered a phalanx of eighty-plus 109s at nearly 30,000 feet southeast of London. The Hurris had a slight advantage in height, but the enemy spotted them a long way off and wheeled away, then broke up into twos, threes and fours, dived hard and disappeared into the thickening haze—very nearly a fog—which covered the Thames Estuary and much of Kent. It was the first time Tuck had seen such a strong force turn tail before a shot had been fired. Jubilant, he brought the wing screaming down in pursuit, shouting to them to split up into their pairs and hunt eastwards through the grayness.

With Carl Capon guarding his tail, he plunged into the haze and down at a few hundred feet spotted the dim silhouette of a 109 dead ahead, going flat-out. They were gaining on him, but too slowly—the mist was getting thicker every moment, and Tuck was afraid that if he waited he'd lose him altogether. He lined up very carefully and scored hits with a short burst. But instantly the 109 swerved—and disappeared.

Tuck cursed and kicked his rudder gently, skidding from side to side, straining his eyes to pick it up again. No luck. But a minute or two later a second victim presented itself, converging from the right and passing ahead at much closer range. Tuck's burst brought forth a big puff of black smoke—then this 109 banked steeply and vanished too.

Intelligence could only credit him with two "probables." Several enemy machines were found to have crashed in the region, but the mist had blanked out the camera-gun films, and the general confusion of the action made it impossible to decide who should be credited with having shot down any one of them.

In his heart, Tuck is convinced that he destroyed that second 109—the big puff of smoke indicated very serious damage to its engine, and since the action was at very low altitude he can't see how the German could possibly have got back across the Channel. Therefore he includes this one in his private, unofficial score.

But on both these big days, as in some other, smaller actions, he had had the distinct and very unpleasant feeling that the third Hurri squadron of his wing were hanging back, letting 257 and John Grandy's boys absorb the first shock of fire in the dangerous business of breaking up an enemy formation. . . .

It was, of course, very difficult to prove that this was deliberate. But the more he thought about it, the more he became convinced that some of these laggards were not using the full power of their Merlins in the battle-climb, despite the order "Buster!" Moreover, there had been an unusual number of cases in which pilots of this squadron turned back because of faulty instruments or engine trouble. And once or twice, out of the corner of his eye he'd seen one or two peel off and drop right out of the party.

He decided to lay his problem before Beamish. The Groupie listened calmly, smoking his stubby pipe. Tiny drops of saliva dribbled from the corner of his mouth and gathered in a glistening blob at his chin, and when he sucked the pipe he made a wet, sputtering noise. When Tuck had finished he said: "So that's the way o' it. . . . Roight then, Robert. I'll think about this." An hour or so later Beamish intimated that all pilots of the wing were to present themselves in the billiards room of the mess at seven.

The pilots were duly assembled and Beamish faced them, his eyes moving slowly, menacingly, from man to man.

"Oi'll not kape you long, gentlemen," he said, in a tone so soft it was almost kindly. "Oi've very good raison to believe that there may be som' among us who're' not as kane as they moight be. Oi'm not a man with an unduly sospicious torn o' mind, but it seems to me that lately we've had a quare amount of engine tro'ble on this station—the sort that's hard to discover, somehow, once the aircraft's com' back and landed. . . . And then, there's this business of too many lads bonching up at the arse-end, and getting into the foight just that bit late. Other peculiar things too, which Oi'll not bore you with at the present.

"Oi just want you to onderstand this: from now on, every toime you go up, Oi'll be following, and Oi'll be watching. And if Oi should see anybody hanging behind or breaking off, and if Oi don't see a clear raison for such a thing, then"—his voice fell almost to a whisper—"Oi promise you faithfully, Oi'll be after him m'self, and do him! And Oi'll do him *entoirly!*" He nodded once, then walked out of the shocked silent room.

After that the third squadron kept up with the others, and for weeks on end no one dared turn back from a patrol with mechanical faults—even though a machine might be vibrating from end to end and the instrument needles quivering on the danger marks. Every time the wing took off and went into its first climbing turn, Tuck, out in the lead, would glance back over his shoulder and see Victor's machine racing across the grass—always the last off. The groupie would shadow them throughout the flight, keeping three or four thousand feet higher and a mile or so behind. Wherever the North Weald wing flew, there was always that last Hurricane added, like a postscript.

Beamish had an adhesive memory. He knew every pilot, airman and N.C.O. on the station by his Christian name. He knew where they came from, and what they'd done in civilian life. He could spot the phoneys, the "flannelers" and twitch cases unerringly, and with some he was ruthless. But with others, seemingly irretrievable, he often performed miracles, displaying surprising tenderness and patience.

On the station he never had more than a very occasional half-pint, but regularly, on a sudden whim, he would drive a bunch of his pilots up to London and lead them on a mad round of pubs and clubs and drink most of them into a fog. Tuck remembers that sometimes on these expeditions, about two in the morning they'd lurch into some dimly lit club "and some terrible old hag would come rushing out of the smoke haze, throw her arms around the Groupie and shriek 'Veector, dar-leeng!' He'd pat her rump and say 'Grand to see you again, Lolla m'dear. Didn't Oi say Oi'd be back? Here's a kiss to prove Oi still love you. Now off with you, and fetch us a drink, loike a good girl.' He had great charm, he never forgot a name, and even the toughest, roughest women thought the world of him. And so did I."

At this stage of his life Beamish never slept more than four hours a night. He didn't walk—he ran, or rather he *bounded* everywhere. Every day he did at least thirty minutes of calisthemics under the guidance of one of the physical training instructors—"to keep m'waist down, so's Oi fit into a Hurri."

In the mid-'thirties, as one of the air force's most promising young pilots, he'd suddenly collapsed with roaring tuberculosis. Doctors told his family he'd never fly again, that he'd be lucky to live more than another three years.

After only a few months in sanatorium Beamish had rebelled, walked out and taken ship to Canada. There, for two years, he worked as a lumberjack in the cold, crisp air of the Far North. On his return an R.A.F. medical board passed him as perfectly fit and he was readmitted to flying duties.

Beamish was an even more profound influence on Tuck than "Daddy" Bouchier had been back at Hornchurch.

"I tried to emulate Victor," Tuck says, "because I considered him one of the best commanders I'd known. While you were under his leadership, you could never do anything without stopping for a second and asking yourself: 'Now is this what Victor says is right— is this what *he'd* do?' He was the kind of man you came to trust completely, the kind of man you tried to be like in every way possible."

★   ★   ★

The hit-and-run Messerschmitts swooped on North Weald repeatedly, but nearly always one of the squadrons got into the air in time to challenge them, or at least to give chase. Landing immediately after one of these attacks, Tuck couldn't find the corporal who was responsible for servicing the camera-guns. Another raiding force was reported on the way, and frantic to get off again at top speed he threatened to have the corporal court martialed if he failed to show up inside a minute. Then Chief Tyrer strode up and said quietly: "Found him, sir. Right over here." He led the fuming Tuck round a small bomb crater and across to the canvas-walled latrines on the other side of the perimeter track. "Having a quiet drag, I suppose!"

"No, sir," said Tyrer. "Take a look."

The corporal was sitting on one of the outhouse seats, stone dead. A bomb splinter had ripped through the canvas, struck the back of his neck and snapped the spine.

Shocked, and in a bad temper with himself now,

Tuck yelled: "For Christ's sake get him off that thing, and pull his pants up! Bloody indecent. . . ."

He took off again without changing the film in his cameras.

Perhaps the only big opportunity missed by Tuck during his whole flying career was the battle of November 11th—Armistice Day. Though furious at the time, afterwards he didn't grudge it, because his friend Cowboy took over the leadership of 257 and scored an overwhelming success.

Tuck had taken the day off on the M.O.'s* advice. He had wakened during the night with a painful eardrum—caused through flying while suffering from a head cold. But by midmorning, feeling much better and bored with his room, he took a shot gun and drove to some nearby woods for a bit of shooting.

Early in the afternoon, as he emerged from the woods thinking about beer and lunch, he saw his wing spiraling up into the sky, forming into three great arrowheads and setting off to the southeast. Now and then one of the Hurricanes became a pure, golden cross as it caught the rays of the watery sun. The spectacle hurt him deeply. He cursed the M.O. who'd persuaded him to stay grounded today, and ran to his car. The little Hillman set the hedgerows aflutter as he hurtled through the country lanes, back to the airfield.

But by the time he got there it was much too late to take off after them. Ops room could only tell him that a big force of "bandits" was approaching the mouth of the Thames and the wing was about to intercept. He didn't want to listen in to the r/t—that would only increase his frustration—so he went outside and slowly paced the tarmac, and after a while Chief Tyrer joined him. Tyrer could see how it was with Tuck, so he didn't try to force a conversation. They walked up and down together, wordless for minutes on end, but neither of them embarrassed or resenting the other's presence. Tyrer was one of the few men

*Medical officer.

with whom Tuck could share that rare and mysterious thing known as companionable silence.

After about half an hour the Hurricanes came roaring home, in twos and threes, their broken gun-ports whistling, wings rocking a little as though tired, some of the engines sounding rough and breathless. Cowboy and Carol Pniak were the first to touch down. Cowboy bobbed out of his cockpit and came over the grass at a frantic, waddling run, waving his arms and emitting a mating moose howl.

"Hey boss—guess what? A mob of bloody Ey-ties!—yeah, Musso's boys! Jeez, what a helluva day for you to be sick!" He caught Tuck's arm and led him towards the parked car. "Quick, boss—in you get. One of them pancaked just a few miles away—over by Woodbridge. Let's take a look at it!"

Pniak joined them in time to dive into the back seat. On the way, between directions, Cowboy gave a full account of the action. Carol, who spoke very little English, lay back and appeared to go to sleep in an instant.

"We were vectored out to sea," Cowboy said, "just off the Thames. Suddenly we spotted nine of the funniest looking kites you ever saw. Bombers—big and fat. Flying slugs—bumbling along in a tight 'vic.' I didn't like to go rushing in baldheaded until I had some clue as to what the hell they were, so I held off and came round alongside them, out of range. Pete Brothers, cool as you like, swanned over, went right up above them and had a good look. All he could say for certain was that they were armed and they sure weren't British—and that seemed good enough for me.

"I led the boys in from the back, line abreast. We started with the rear starboard kite and sawed right across the formation, finishing up with the rear port boy. It was then, when we got in close, that I saw the Ey-tie markings.

"Listen, Bobbie—they're not such chickens, these guys! They stood up to it bloody well. Badly outnumbered, too. Their gunners fought us all the way, and that formation didn't even bend. I saw they were going into a climb—striking up for some thick cloud around 20,000. I knew we had to break them up before they reached that, or we'd lose them.

"At the second pass we did it. Two of them were badly shot up and when they dropped out the others started turning in all directions. I singled one out and followed him round. I gave him a burst, and damned if he didn't pull his nose straight up and go into a loop—honest!

"I thought he was trying to fox me—trying to make me break off. I've never seen a bomber do anything so violent. Right up on his back, he was. I thought to myself the crew must be rattling around inside there like peas—unless, of course, these wallahs strap themselves into their seats.

"Anyway, I stuck to him, right on round, and he fell off the top of the loop into a vertical dive. I followed, waiting for him to start pulling out. But he didn't. One 'chute came out, and then the whole issue started to crumple like a wet newspaper. Next thing I knew it suddenly burst into hundreds of small pieces. They fell down to the sea like a bloody snowstorm.

"I reckon I know what happened. My burst must have killed the pilot. I think he fell back, pulling the stick with him—that's what caused the loop, see? Then, when the kite fell off the top of the loop he'd slump forward again, and his weight would hold her in a vertical dive. She kept on building up speed until she broke up.

"I climbed back to look for the others, and found two bombers, sticking together, covering each other.

"Both seemed to have been badly shot up—streaming smoke and oil. I was just getting set to have a go at them when right out of nowhere a third bomber came straight at me in a dive, flaming like a torch. I was sure he was going to hit me, and I whipped the stick over. I tell you, Bobbie, I *felt* the heat of that thing going by!

"He plunged right down into the water. I was pretty shaken. By the time I'd stopped twitching, there was no sign of the couple I'd been going after.

"I reckoned the party was over, and now I could go home, but as I was turning I saw a dogfight right above me, between our boys and another queer-looking bunch—fighters now, *bi*planes, for Christ's sake! I'm sure they were Fiats—I've seen pictures somewhere, maybe in the recognition books. Between twenty and thirty of them, camouflaged in brilliant colors. Real pretty, they were—but they were full of fancy tricks.

"I went up to join in, but the one I singled out saw me coming and just as I got on his tail he whipped round in a quick turn. What a sweet little kite that Fiat is—maneuverable as a taxicab! All the same, I got in one or two bursts.

"Boy, that was the longest dogfight I've ever had— tight turns, climbing turns, half-rolls—all the flashy stuff. That Italian could certainly fly. I thought we'd never stop—every time I got him in my sight he'd flick over his back and do something quite unexpected.

"Neither of us seemed to be getting anywhere until one of my bursts caught him amidships, and then for just a moment he looked completely out of control. But would you beat it?—suddenly he did something like an Immelmann turn and came in at me, head-on. I went into a diving turn and we started this here-we-go-round-the-bloody-mulberry-bush stuff all over again.

"I got in two or three more bursts and this time knocked some fair-sized chunks out of his wings and fuselage. Then my ammo gave out. God, I was mad!"

"Rotten break, Pete," Tuck sympathized. "All the same, you put up a marvelous show. Lucky sod!"

The Canadian gave him a big grin. "Hold your horses, Bobbie—there's more to come. I couldn't start for home, because I was scared he'd fasten on my tail the moment I broke off. So we kept up this turning and twisting routine and suddenly—more by luck than judgment—I found myself bang-on his tail, only about thirty yards behind and a few feet higher. I could see the pilot turning his head and staring back but I couldn't make out his features—just a white splotch under the goggles. If I'd had even a dozen bullets left I could have finished him off easily. It was enough to make a nun curse. . . .

"I guess I went kinda haywire. He was so near, and the damned thing looked so light, I thought I'd take a chance and try and knock him out of the air. So——"

"You *what?*" Tuck shot him a glance, saw by his expression that he wasn't joking.

The Canadian rubbed his jaw and shrugged: "Well- . . . the goddamn thing looked like it was made of boxwood and string, a mere toy. Crazy, I know, yet at the time I honestly couldn't see how it could damage a solid job like the Hurri. But as I started to close with him I had second thoughts, and decided I'd just try scaring the living daylights out of him instead.

"I aimed for the center of his mainplane, did a quick little dive and pulled out at the last second. The idea was to pass very close over his nut and maybe send him into a spin. But I must have misjudged it. There was a slight bump and a bit of a shudder. I think I hit him after all."

"You *think*—you raving lunatic, you mean you're not sure?" Tuck was astounded by all this; he'd known Cowboy had strong nerves, but he would never have thought him capable of such fanaticism! One of those balmy Poles, maybe, but not this plump, easygoing Canadian. . . .

"Yeah, I'm pretty sure. I climbed and circled, but I never saw him again. The Hurri was vibrating, and the engine revs wouldn't stay steady. Six inches were missing from one prop blade, and nine from another—of course I didn't find that out until I came in to land.

"By now another squadron had arrived on the scene and they were chasing the Fiats all over the sky. I heard their leader on the r/t and recognized his voice. It was Elmer Gaunce, with Number 46. Elmer's an old pal of mine—knew him back in Edmonton, Alberta, when we were both kids.

"I cleared out and headed for home, but on the way I saw a Hurri, just below, having the same kind of affair with a Fiat as I'd just had." He twisted and looked into the back seat. "It was old Carol, here—look at him, will you, out like a light! I guessed he must have run out of ammo, too, so I went down and did a dummy head-on attack on the Ey-tie. At around two hundred yards he wheeled away and pelted out to sea.

"Carol and I continued on our way, and before long we saw another Hurri with three Fiats worrying him. I dropped down and feinted another head-on attack, and again at about two hundred yards the Fiats broke off and headed east. And that really was the end of it—except, of course, that just as we were entering the circuit we saw this bomber being chased inland by two of our boys and crash-landing out here. Can't be far now—turn left at the next crossroads and we ought to be there. . . . Hey Carol, wakey-wakey!"

A few minutes later they found the Italian, crumpled against a thicket of firs not far from the road. Tuck—who took his aircraft recognition seriously, and hadn't confined his studies to German machines like most of the others—identified it as a B.R.20. It was riddled with bullet holes and badly smashed up.

Some local people and two policemen were already on the scene, and had brought out most of the crew. Three of the Italians were conscious, but seriously injured. One of them had his right forearm almost completely severed and was bleeding copiously.

Cowboy grabbed one of the constables. "What's the matter with you, man? Get a tourniquet on him, quick!" The policeman looked dazed and sick, but he nodded and went to work. As the tourniquet was tightened the Italian made no sound, but the pain and the fear shrieked in his eyes. Then he turned his head and caught sight of the three air force uniforms. His gray lips moved feebly once or twice, and they caught the word *"Pilota."* Then he passed out.

Cowboy led the way to the wreck, not looking at the other still forms stretched out on the grass. The side door of the bomber was huge—it seemed you could have driven a car through it. As he approached it Cowboy trod on what looked like an old piece of sacking lying on the ground. His foot sank into something soft, one end of the lump rose about a foot and

there was a loud, vulgar noise—unmistakably a belch!
Cowboy sprang back, his body arched like a cat's, and
cannoned into Carol and Tuck. Now, on closer inspection, they could make out the vague hump of a body
under the sacking.

"Christ—it sat up! The poor bastard's still alive!"

"No," Tuck said quickly. "You must have stood
on his stomach, that's all. Gases, you know. . . ." And
then he gave a short, hard laugh, because it seemed the
only thing to do. Cowboy shook himself, grinned
sheepishly, then picked his way round the obstacle and
clambered into the aircraft.

Just inside the door the top-gunner was still in his
harness, swinging gently, full of bullets. The harness
creaked faintly and the floor beneath him was slippery.
They had to flatten themselves against the side of the
fuselage and wriggle past. On the way Tuck looked
up and saw a holster at the gunner's waist. He
reached up, extracted a Beretta automatic and stuck
it in his pocket. Since ten years of age he'd been a keen
collector of firearms, he still couldn't resist a chance to
augment his private armory. . . .

In the waist they found two hampers, large as laun-
dry baskets. One was stuffed with a variety of foods—
whole cheeses, salami, huge loaves, cake, sausages
and several kinds of fruit. The other held still more
food, and over a dozen straw-jacketed bottles—Chianti.

Cowboy whistled softly. "They sure do it in style.
What d'you think, boss? Pity if all this went to waste
—rotting in some cubbyhole at Air Ministry." Tuck
nodded, and between them they carried the hampers
to the car. One of the policemen watched them, frowning, but Tuck went to him and explained in a purposely
offhand manner: "I'm the commanding officer of 257
Squadron, North Weald. My chaps shot this one down.
We're taking a few souvenirs—that's an unwritten law
in the R.A.F. If anybody raises any objections, refer
them to me." The constable rubbed his collection of
chins, but said nothing.

They went back to the wreck and picked up several
steel helmets and bayonets. Finally, just before the

troops and the recovery trucks arrived, they took the bayonets, climbed up on to the twin tail unit and cut from the tip of each fin the beautifully handpainted crests: they'd look splendid behind the bar at the mess.

Back at the airfield, a flurry of photographers greeted them. The "Burma" boys posed with their trophies, and the Chianti bottles circulated merrily.

Next day the papers blazed with highly-colored reports of the battle. It was with difficulty that the 257 pilots sifted out the new hard facts.

Mussolini, it seemed, had stated in a broadcast that he had asked his friend Adolf Hitler for the "privilege" of participating in the airwar against Great Britain, and had received consent for a few crack squadrons of the Regia Aeronautica to use Nazi bases in France and operate "side by side with the Luftwaffe."

But the Germans had kept well clear of their slower, inexperienced allies. While the Italian mixed formation headed for the Thames Estuary, a German fighter wing had attacked a number of towns along the southeast coast. Other British squadrons had opposed this "diversion," destroying twelve of them for the loss of two.

The Italian force was estimated to have consisted of nine B.R. 20 bombers, with an escort of between thirty and forty of the nimble Fiat C.R. 42 fighters. The Regia Aeronautica had lost at least twelve machines —of which 257 bagged five, or perhaps six—without destroying a single British fighter, and entirely failed in its mission, which was to bomb shipping in the Thames.

The newspapers made fun of the Italians, making out that they'd fled at first sight of opposition. This greatly annoyed the North Weald pilots, who thought the Ey-ties had displayed plenty of courage—especially in view of the inferior performance and speed of their machines.

*   *   *

London was taking a terrible beating by night. So were other great cities—Glasgow, Birmingham, Manchester, Coventry and the ports. The pilots who'd stopped the enemy by day throughout the summer and autumn were depressed and angered by the fact that Hitler was now achieving serious destruction with "the Blitz." It seemed that all their efforts had been in vain.

Circumstances demanded that the day-fighters should make some attempt at continuing their activities into the night—but "attempt" was all it could be called. Long hours of fumbling about in the blackness . . . often hopelessly lost . . . sometimes being caught in their own searchlights, or fired at by the A.A. defenses —even by their own overzealous comrades. And in those first weeks the results were negligible.

The specially trained and equipped night-fighters did a little better, but they were still far too few, and the bulk of the opposition was supplied by the guns. Still, very gradually the effectiveness of radar apparatus and fire control methods improved. Decoy fires, dummy airfields, devices which bent the raiders' direction-finding radio beams—all these began to reduce the Germans' striking power. If they could not destroy the nocturnal invader, at least they could divert him from his targets. . . .

The men who had been trained to fight, to face the enemy fire and all the other hazards and ordeals of war, stood amazed at the spirit of the civilians. No mass exodus from the cities. No big fall in production. No wavering, no serious complaining, no blaming the air force. Every night thousands of people were dying, buildings disintegrated, communications were cut; but still each day the munitions factories throbbed and the trains ran and the shops provided fair shares for all.

The Royal Family stayed on in their home in the very heart of the capital, a huge target for the bombers, and one they could hardly fail to identify. It was said that the King had been asked by the government to leave, but had refused. All this steadiness and solidarity greatly impressed the flying men, strengthened their resolution and increased their hopes.

On one of the few nights that Tuck and Cowboy had up in London during this period, they came out of

Shepherd's with two pretty A.T.S.* officers in the middle of a raid. After one particularly heavy explosion they saw an aged taxidriver with a walrus mustache stop his cab, run to the middle of the road and wave his fist at the dark and hostile sky. In sulphurous Cockney this old man told the cream of Germany to go and do some rather awkward things. He was completely oblivious to the young people watching him—this was a purely private conversation, and he ended it with a very eloquent raspberry that made his droopy mustache flutter. Then he hobbled back to his cab and drove off with a fierce grinding of gears.

The two girls were rather shocked, but their escorts howled with delight. Nothing could have typified so gloriously the earthy defiance of the ordinary British people. And the incident brought home to them just how things had changed since the summer fighting—*then* they had felt remote from the civilian populace, *now* they felt they were bound together by common experience and a common anger.

* * *

He never took more than a day or two of leave, and he hardly ever went home now. It wasn't that he lacked filial affection—in particular, he remained devoted to his mother—but somehow the place got terribly on his nerves.

He kept in touch with the family by telephone, talking with his father for ten or fifteen minutes about once a week. Mrs. Stanford Tuck wrote with great regularity —cheerful, chatty letters, never emotional or too personal. He was still managing to reply to these occasionally, and he never forgot to send her flowers or some little gift on her birthday or wedding anniversary.

Under the circumstances, it was a sensible and dignified relationship. Perhaps the very fact that his parents seldom saw him kept the family calm and hopeful.

* * *

Franek Surma bailed out at a few hundred feet after a 109 had bounced him on take-off. His 'chute caught

*Auxiliary Territorial Service.

in some trees and he finished up dangling ten feet or so above the ground at the back of a local pub. He had two gorgeous black eyes, and was picking pieces of the instrument panel out of his face. Lunch time drinkers rushed out and heard him cursing in a strange language.

As it happened, Franek always wore a rather flashy Luftwaffe flying jacket which he'd whipped from a wrecked bomber back in the Polish campaign, and he hadn't even bothered to remove the German eagle insignia. Among the people from the pub were a large group of Free French soldiers who wanted to lynch *le sale Boche* right there in the tree by the simple expedient of climbing up and winding the shroud lines once or twice around his neck.

While the landlord and local people argued with the Frenchmen, somebody had the presence of mind to phone the airfield, which was much closer than the nearest police station. Tuck, Cowboy and "Buster" Brown, the adjutant, drove over at breakneck speed in time to rescue Franek. Once everything had been explained to the French, they poured out apologies, lifted the Pole down tenderly and rained kisses on his blood-streaked cheeks. And when a doctor had been found to dress his wounds, they refused to let his comrades take him back to the 'drome until everyone concerned had joined them in the saloon for several rousing toasts in furtherance of the Allied cause. Franek finished up roaring patriotic songs, passionately embracing the Frenchmen—and several pretty girls who chanced to pass near. He seemed to get plastered very quickly, rocking to and fro precariously. It wasn't till late that afternoon, when they got him back to sick quarters, that they learned he was suffering from severe concussion.

After this incident Tuck prevailed upon him to cut the Nazi badges from his flying clothes.

* * *

Sir Hugh Dowding left Fighter Command at the end of November. His qualities of restraint, determination, cold computation and canny husbandry, and his policy of holding back the Spitfires for home defense during the Battle of France, had triumphed gloriously.

But now, with the period of acute peril over, and with the supply of new pilots from the wartime training scheme mounting steadily, "Stuffy" Dowding's careful, calculating methods were hotly criticized by other commanders and members of the Air Staff. There was a growing school of thought which maintained that the best—some of them said the only—way for Fighter Command to counteract the Luftwaffe's night-bombing was to "stretch out" and strike at the Germans in France, Belgium and Holland.

So this quiet and dignified man, his great ability and his lustrous achievements acknowledged by all—even by those who, with the change of circumstances, had become his fiercest critics—quit his place in the forefront of the air war and went on a mission to the United States, where he justly received a tremendous welcome.

His successor was the handsome and popular Air Marshal W. Sholto Douglas, a dark dumpy, smoldering man. At 47, he had served as Deputy Chief of Air Staff and was reckoned a tough boss and a brilliant strategist. He seemed always to be tensed up, exercising great self-restraint, assuming a very soft voice and a quick, mild smile in order to control seething impatience.

Surprisingly enough, Sholto Douglas did not immediately embark on a program of sweeping changes. The R.A.F. was still seriously outnumbered by the Luftwaffe, and could not yet risk heavy losses. The enemy would be quick to sense any weakness in the defensive organization, and to thrust against the gaps.

By day the Spits and Hurris now were the masters. But once the sun dropped below the western horizon, the marauders worked their evil at comparatively small peril. The most urgent task, therefore, was to build up the night-fighter force, and to increase its efficiency. Accordingly, experienced men from the day squadrons were given aptitude tests, and those who proved suitable were whipped off to "night school." For a while this problem, and others relating to supply and administration, appeared to preoccupy the new C.-in-C., but all the time a part of his mind was fixed on the future, shaping a bold new policy which would enable swarms of his fighters to roar eastwards day after day and fall, like avenging angels, upon the enemy bomber squadrons who tortured Britain by night.

Very early one wintry morning a conference was held in Sholto Douglas's office at Uxbridge, Middlesex. Several officers were summoned, each of them an active pilot and experienced air leader, and they included Bader, Malan and Tuck. The question put to them by the C.-in-C. was, in essence: *should the fighters be armed with the new 20 millimetre cannon, or should they stick to the Browning .303 machine guns?*

Bader, who had left 222 months ago to lead a Canadian squadron, was the first to answer.

"Not the slightest doubt in my mind," he declared right off. "Stick to the Browning, sir. Damned fine gun. Served us well enough this far—chaps know it outside in, pilots *and* armorers." He raced on, in a confident, grunty voice—listening to him, Tuck had to fight to stop himself swearing aloud. Bader spoke as if the whole thing were childishly simple, and he took it for granted that everybody would agree with him—he actually seemed a little amused that he should even be asked to state the obvious like this!

Tuck wondered why it should be that he and this man—this very remarkable man, with all the guts and ability and so many other admirable qualities— seemed destined always to rile one another. Fortune had teased them into a strange rivalry, and this was one of the times when it could very easily grow bitter.

Yet they *could* get along together—they'd proved that more than once over the last two months on the odd occasions when they'd met at some airfield or London pub. So long as the talk was light and unimportant, they made happy enough company—but whenever "shop" monopolized the conversation, very soon they'd fall out, because then each seemed to possess an infallible knack of saying the one thing that would ridicule the other's pet theory of the moment.

Now Bader was trying to destroy Tuck's long-cherished dream, to deprive him of his beloved cannon. Bader, with his big gestures and aggressive garrulity, his amused superiority and his startling, gleaming eyes. . . . Tuck jammed a fresh cigarette between his tight lips and held his silence, taking long, slow breaths through his nose, feeling a red flag of fury unfurling on each cheek.

Quite recently everyone in Fighter Command had heard details of the experiments which Georgie Proudman of 65 Squadron had carried out with cannon a few weeks before he was killed. Georgie had been detached to fly a Spitfire armed with two 20 mm.s both on target trials and in actual combat with enemy fighters and bombers.

The cannon had been mounted on their sides at that stage, and this had given rise to some serious faults—because friction was imposed on certain moving surfaces which were never intended to withstand it. Nevertheless, so long as the cannon were fully serviceable, the results had been impressive and the pilot's report on completion of the tests had been most enthusiastic. Tuck knew more about these trials than most people, because he'd spent a whole day with Proudman and questioned him ravenously.

Now it was understood that the boffins had found a means of mounting four cannon, right side up, in the wings of a fighter, and that range tests with these had been appreciably more successful than Proudman's. Tuck had rejoiced to think that soon there would be delivered to his hand a new and wondrous weapon. . . .

". . . any big change over would do more harm than good," Bader was saying. "The Browning's a proven weapon. Damn sight sooner carry on with eight of them than start monkeying about with four—or six, hell, even eight—of *these* things. Don't trust cannon—too bloody new, probably still full of bugs. Let's wait till they've worked out a few modifications."

The rasping of a match. Bader was lighting his pipe,

signifying that he rested his case. The C.-in-C. turned to Malan—"Sailor . . . ?"

Malan of the mild exterior, the ruthless destroyer with the smiling eyes and the patient droop to his shoulders, ordinarily wasn't a great talker. But today he expressed himself forcibly and at considerable length —and his views chimed melodically with Tuck's:

Nowadays the enemy used bags of armored plating . . . at anything over five hundred yards Browning fire was unlikely to inflict serious damage even with a high percentage of hits . . . more and more interceptions were being made in darkness, or, in the case of lone raiders, cloudy conditions, so the attacking pilot often had only time for a two or three second burst from long range . . . a single 20 mm. shell, even at eight or nine hundred yards, could result in complete destruction of an aircraft . . . if they were given a mixture of explosive and armor-piercing shells, results were bound to improve, and moreover (here a point that brought a quick nod from Sholto Douglas), the fighters would be well-equipped to deal with enemy ground installations and shipping. . . . Jerry had been using some cannon from the start, and even some of the French fighters had been armed with them—there was nothing very original or "revolutionary" about the idea. . . .

While Sailor spoke, Bader chewed on his pipe and shifted his thick body angrily in the chair. As ever, he hadn't expected anyone to disagree. Tuck could feel the South African's words rebounding from the wall around the man.

"Tommy—let's have your ideas."

He hesitated, reining in his fervor, and then he simply said—with that artificial, over-precise intonation that always crept into his voice at difficult moments like this—"I agree with Sailor, sir. Absolutely. I've nothing to add—except that if you give us cannon I'm quite certain we'll knock down a lot more of them than we do now."

Immediately, Bader gave a snort and said: "Bobbie, for God's sake don't talk rot! You too, Sailor—the pair of you think this is a bloody miracle weapon, and the damned thing hasn't even been——"

Tuck couldn't contain his agitation an instant longer. He sprang to his feet, words foaming—words that he didn't have time to check out, and that he couldn't remember afterwards. For perhaps half a minute he and Douglas were both shouting and gesticulating—Bader, with his incredible agility, swiftly rising on to his metal legs to meet the challenge. Then the C.-in-C. raised a hand, like a bishop about to bestow a blessing, and said: "Now, now gentlemen . . . no use howling at each other, that won't help me."

They sank back into their chairs, each a little surprised at his own behavior. Sailor grinned and winked at each in turn. Then Sholto Douglas asked the whole group a number of questions, and the conference ended without further friction.

As they walked to their cars, Tuck swallowed hard, summoned up a muscular smile and said to Bader: "Sorry I blew up in there, Dougie."

"We both did."

"Yes . . . we always seem to get on each other's nerves. Funny, isn't it?"

"Aw, beetle off." A sudden, friendly enough grin.

"You too, you old sod."

Bader lurched off to his car, while Tuck and Sailor lingered on the path. Just before he got behind the wheel, Bader turned and called back: "But you're wrong, y'know. You're both bloody well wrong!" And then he was gone in a haze of blue exhaust smoke.

"Exit the Demon King," said Sailor, eyeing the lingering fumes. They walked slowly to Sailor's car and he got in and started up. Tuck sighed and shook his head.

"Christ, he makes me so mad. Why does he have to be so obstinate all the time?"

"Because," Sailor said as he moved off, "if he wasn't so bloody obstinate, he bloody well wouldn't be here!"

*   *   *

Not long after that a Messerschmitt fighter-bomber nabbed the North Weald Wing taking off in formation. Tuck was leading the first "vic" of three, and they'd hardly got their wheels up when one of the bombs exploded just off the end of the runway and a shade to the right. A vortex of flame and dirt reared up almost directly under their wings.

They were only about forty feet off the deck. Tuck's aircraft ballooned crazily and the starboard wing whipped up. By swift manipulation of the stick and rudder he managed to fight her back on to an even keel. But Jock Girdwood, on the right, had caught the full fury of the blast. Tuck jerked his head up and saw Jock right above him, scant feet away, upside down!

He could see right into Jock's cockpit—and in that icy instant, through the stormy flood of the years, flashed the memory of Sergeant Gaskell's hands frantically working the controls of his doomed Gladiator. . . . Rigid, pressing himself back in the seat, he waited for the agonizing grind and roar of collision. But Girdwood's Hurricane slid over him, across to the left, completing a half roll. Then its nose dropped and it dived into the ground and exploded with a glare like the sun's. It had all happened in a matter of two or three seconds.

*　　*　　*

More "kills" in the logbook:

December 9, over sea.	Whilst flying over sea sighted and chased Do. 17 or 17z. Shot it down 10 mls. N. of OSTEND. Almost dark.
December 12, over London.	Ran into about 40 Me. 109. Squirted at three then singled out one. Chased him right out to sea and shot him down in water off Clacton.

And then . . .

December 17, to Coltishall.	Squadron moved back for "rest cure." Perfectly bloody.

He needn't have worried—they got no rest. Droves of German fighters and fighter-bombers and an occasional lone raider continued to breach the wintry walls of cloud over the Channel and it was found that 257 couldn't be spared after all. And as it turned out, the move to Coltishall in Norfolk—was another of those oddly disguised blessings. For within a few days it brought into his life someone who was to prod his neglected spirit into wakefulness, and bring him back into touch with the big, permanent world outside the airfield's boundary wire.

Someone named Joyce.

13

She was standing on one side of the old-fashioned hearth, holding a dainty glass in one hand and a cigarette in the other. A tall, strong, straight-backed girl with cool, agate eyes and abundant, honey-blonde hair that seemed to stir like smoke each time she turned her face to one or other of the people clustered round her.

Tuck saw her the instant he entered the room, and he stopped dead and stood staring for a long time from just inside the doorway, across the breadth of the dance floor, lifting and weaving his head whenever a fox-trotting couple blocked his view. Not for years had he experienced such a powerful emotional surge. It was totally beyond his understanding and control. A fierce, sweet pain in his chest—caused simply by looking at a strange young woman? It was the most ridiculous and mysterious thing that had ever happened to him, and his instinctive reaction was one of resentment. He struggled to be objective, to fight off this embarrassing and unnerving sensation to reason it away——

Firstly, she's smoking. . . . (He'd always thought it an inelegant habit in women.) She smokes hungrily, like a man. She takes big lungfuls of it, and exhales in powerful little jets, through her nostrils as well as her lips. See how pale she is—*unnaturally* pale. Not the outdoor type! Not beautiful, you couldn't say that. And her expression . . . it suggests a polite tolerance— faint amusement, covering up boredom. . . . Very superior type, good family, finishing school and all that. . . . And that dress, hideous color! (It was royal blue, which had always revolted him—and still does. Joyce had made the dress herself. She never wore it again after that night.) Seems too tight. Sleeves are a funny length, neither one thing nor t'other. Definitely a weird getup. . . .

But for every fault he sought out, and strove to enlarge, an arguing voice within him listed two or three assets: the cool grace of every little movement of hands or head—the clear, white texture of her skin, startling even at this distance—the softness and the thrust of her body under that tight dress—the poise and the strength and the overwhelming femininity. . . .

Then suddenly, as he stood staring across at her, she half-turned, laughing lightly at some joke, and there came a momentous instant when their eyes met . . . locked . . . seemed to speak, to say simply: "Ah—so *here* you are!" It was as if they had clasped hands. He found himself moving round the dance floor, through the noise and the smoke towards her.

He left behind him three baffled companions: Cowboy, David Coke and their new station commander Group Captain W. K. Beisiegel. The Groupie had persuaded the three 257 pilots to come along here tonight, to this unpretentious little dance in the upstairs bar of the "King's Arms" at North Walsham which the good-hearted local people held once or twice a month for officers from surrounding army and R.A.F. bases. Not surprisingly, "Bicycle" was a bit put out at Tuck's disappearance before he'd had a chance to make introductions to the organizers. Cowboy and David had no chance to think up any excuses for their leader, because they were rendered utterly speechless the very next moment by the spectacle of him sweeping a comely girl on to the floor and into a brisk quickstep—apparently quite at ease, talking and laughing, and making the girl laugh too.

A part of Tuck was just as flabbergasted as the watchers. Except for the foursomes organized by Cowboy, he'd avoided girls at all costs up to this moment.

Life for him had been only battles, to win or to lose, and to start again each dawning. So, like a priest, he had cast off all stirrings and yearning by denouncing them as temptations which would weaken his faith. The flying war had been his oyster and now, as he felt the temptation to emerge at last, despite his outward gaiety and assurance, he was not without fear.

And Joyce—how did she feel that night?

"I was instantly very attracted," she says. "In fact, much too attracted to remember what we talked about during that first dance together. I don't think I really heard what he was saying.

"I thought he was quite wonderful—tall, slim and very dashing, with his dark mustache and his beautifully cut uniform. But later, when we left the floor and started getting to know one another, I was rather shaken when he told me he was just twenty-four. There were deep lines on his face and circles under his eyes. I wouldn't have thought he was a day under thirty."

They were together for the rest of the evening. Conversation flowed with superb ease—Tuck had never talked so freely and naturally with any female.

Joyce lived with her parents in a large house a mile or two from Coltishall airfield. She was on leave from London, where she'd been serving as an ambulance driver. Neighbors had invited her here tonight and she'd come purely out of politeness, not intending to stay more than an hour or so. She'd borrowed her mother's Morris Ten, and she told him she'd be glad to give him and his friends a lift back to the airdrome.

"Wizard," he said, "but I think the others have their own transport. If you'll excuse me a moment, I'll see how they're fixed." He knew exactly how they were fixed, because he'd brought the three of them here in his own service car. He found Cowboy and gave him the keys. "Take the others back for me, Pete. I'm organized."

"Sure, boss." Cowboy glanced across to where the tall blonde girl was sitting, his tongue swelling his cheek, laughter flickering in his eyes. "What'll I tell the Groupie? He's a bit drunk, y'know."

"Oh, tell him . . . tell him I ran into a very old friend."

★   ★   ★

Joyce's parents had gone to bed. They sat by the fire in the big, softly lit lounge with the wind pressing on the window panes and talked into the wee small hours. They talked about the most unlikely, quite inconsequential things—dogs, swimming, cars, wines, foreign foods, tennis, holiday resorts, photography, horse riding, Tommy Handley, market gardening, Wilfred Pickles, skiing. . . . They did not talk about the war. She didn't ask him about his flying, and for once he seemed quite able to find other topics.

He was still a little afraid of what was happening to him. He had the uneasy feeling that he was just discovering something that other men found much earlier in life.

A streaky dawn was in the sky when at last they parted by the airfield gates. He stood with his hands on the car door, smiling down at her. There had been no passion between them, but something so much more important that neither had even hinted at it. But each knew that the other knew.

"Tonight?"

"Yes, Robert." He loved the way she said his name.

"I'll phone. About six, probably."

"That'll be fine. Good-by."

"Good-by, Joyce." He stood back as she deftly reversed the car in the narrow road, turned it, and sped off.

*   *   *

Tuck appeared the all-or-nothing-courage fanatic; nobody suspected that nowadays there were odd moments—as the dispersal phone bleated, or while he was buckling his straps in his cockpit—when things didn't seem so good, when his hands shook and his throat tightened and all he could see before him was Joyce's face in the soft firelight. He wasn't fool enough, just because of the woman who'd broken into his life, to start telling himself that he was going to survive after all: he remained resigned to meeting his death in the air, probably quite soon now. Nevertheless, Joyce made a difference: for the first time in his fighting ca-

reer he began to experience these twinges—and they
were only twinges—of regret.

No doubt there were some who noticed that his
laughter was a little too loud, and lasted a little too
long, to be wholly sincere. It was true that he
couldn't stay still for more than a few seconds—in the
mess he'd sometimes spin round in the middle of a
conversation and take a kick at an imaginary football,
and he was always glancing out of the window, or at
the wall clock, or his wrist watch—but such things
weren't all that unusual in Fighter Command at the
end of 1940. Almost everybody, to some degree, was
obsessed with that strange need to hurry—to hurry
with his work, his play, his friendships. His life.

He still had his burning sense of duty, his savage
aggressiveness and those supple and adept hands
that thought for themselves. But wilier now, more calculating, not quite so wanton or impulsive. Better
than ever, in fact, for this tempering of his zeal ma-
tured him into a more brilliant air leader.

Sometimes it was hard to see how he had reached
certain conclusions—how he could have known that a
situation was going to develop the way it did. There is
some sixth sense that a man acquires when he has
peered often enough out of a perspex capsule into a
hostile sky—hunches that come to him, sudden and
compelling, enabling him to read signs that others
don't even see. Such a man can extract more from a
faint tangle of condensation trails, or a distant flitting
dot, than he has any reason or right to do.

The "rest cure" days flamed and roared by, with the
big Messerschmitt packs splintering out of the wind
carved clouds like schools of flying fish from a heaving
sea, and the lone Heinkels and Dorniers, those evil
caravels, restlessly, ceaselessly prowling and probing in
the murk.

They flew patrols from before dawn until long after
dusk, playing a relentless and frustrating game of hide-
and-seek. But now and then their perseverance was rewarded by the destruction of an intruder, and the
squadron's score mounted steadily.

By this time he had begun to tell Joyce some of the squadron's doings. At first it was only the funny things —the narrow scrapes some of the others had, the successes, and the daft mistakes that *could* have been serious. She listened attentively, without a single exclamation of alarm. Her eyes never showed a sign of womanish fears, and when he finished she'd usually just laugh—a warm, wonderfully genuine laugh. But if he didn't want to talk about his work he didn't have to—she never questioned him.

All the same, in other matters he seemed to enjoy small mystifications. For instance, if he turned up late and she made the mistake of asking what had delayed him, invariably he'd wrinkle his nose and eyes and just say *"A-ha!"* It could have been irritating, but she simply stopped asking even such trivial questions.

Joyce didn't pretend to know him—nobody really knew him. He was too complex. He didn't know himself.

One by one she met his pilots, and with them she was quietly gay and didn't try to be "understanding." Soon he began to tell her pretty well everything they did in the air, even about the toughest fights and the losses. She was very calm and composed over the worst of news. If she worried about him, she never let him know it, and when he saw that he didn't have to make the slightest concession to her, his last lingering doubts about their relationship—this strange new adventure— vanished. Though he didn't face the fact, he was completely, contentedly, in love.

They couldn't be together on Christmas Day, because all morning the Coltishall squadrons were at readiness and in the afternoon they flew some abortive patrols. And in the evening the service tradition of the officers and N.C.O.s waiting upon the men at dinner table was observed as usual—and then of course there was a general party in which rank was eliminated and the airmen exchanged tunics with the pilots. L.A.C.* Hillman made a weird Cinderella, having made his swop with the C.O.—Tuck's tunic came down halfway

---

*Lance Air Corporal.

to his knees, but it couldn't be buttoned across the Cockney's bulky chest.

"Cor, ain't you skinny!" cried the L.A.C. whose tonsils were by this time well afloat. "Your Mum never fed you up right when you was a nipper, sir, that's what. You badly wants building up, you do, and that's the God's truth and excuse me for sayin' so. 'Ere —what rotten sod's whipped me wallop? I put it dahn right there. . . ."

"But Hillman, I eat more than anybody else in the mess. I'm always first to call for 'seconds.' Ask anyone."

"Eh? Well then, all I can say is you've got bloody worms!" And with that Squadron Leader Hillman departed, very belligerent, in search of his stolen beer. For weeks afterwards, Tuck kept reminding him of his diagnosis: whenever the Salvation Army's mobile canteen rolled on to the tarmac he'd toss him a shilling and yell: "Get me a cuppa, Hillman—and something solid for the worms." And the L.A.C. would blush like a maiden.

On December 29th at about 11:20 a.m. the radar screens picked up a lone Dornier 17 speeding southwards down the Norfolk coast at exactly 4,000 feet, doubtless photographing towns, harbors and shipping. A light mist lay over the land, but above the water the air was crystal clear.

Tuck and Carl Capon were sent up and vectored on to the prowler. Heading eastwards through the mist, following control's directions, they had to fly with the greatest precision in order to make the interception "spot on." For if the fast, stripped down recce-plane spotted them from any distance, it would turn away, put its nose down and have a very good chance of reaching the French coast before the Hurris could catch up. Tuck's plan was to let control guide them to their quarry under cover of the mist, then they'd pop out and have him in range immediately.

An error of two miles an hour or a degree off course could wreck the scheme. They riveted their eyes on their instruments and made tiny, fiddly adjustments every few seconds. And sure enough just as the controller reported: "Plots coinciding—*now!*" they burst into the clear air and saw their enemy converging swiftly, a matter of five hundred yards distant. Only brilliant pilots could have done it.

The raider had no chance to turn away. Carl lunged straight for it in a beam attack, and Tuck swooped round on to its tail. Once Carl had passed over the bomber and got out of the way—having punched some holes in its withers—Tuck opened up from a hundred and fifty yards. At once beads of oil appeared on his windscreen. Another very long, steady burst chawed off a large piece of the tail and started the port engine smoking. The Dornier rose suddenly, violently, in a salmon-like leap. He closed to a hundred yards and belted it again, up into the smooth, blind belly. Then over and down it went and smacked into the sea just off Great Yarmouth, without having fired a shot. They circled the patch of foaming water, but saw no survivors.

New Year's Eve he'd intended to spend at Joyce's home—quietly, because he could never see the point of a wild celebration just because the world was another 365 days older. But in the afternoon a signal came over the teleprinter from Headquarters 12 Group:

H.M. THE KING ON RECOMMENDATION OF THE A.O.C.-IN-C. HAS BEEN GRACIOUSLY PLEASED TO APPROVE AWARD OF DISTINGUISHED SERVICE ORDER TO SQUADRON LEADER R. R. S. TUCK, D.F.C. AND BAR SQ.257 STOP OFFICIAL CITATION STATES QUOTE THIS OFFICER HAS COMMANDED HIS SQUADRON WITH GREAT SUCCESS AND HIS OUTSTANDING LEADERSHIP COURAGE AND SKILL HAVE BEEN REFLECTED IN ITS HIGH MORALE AND EFFICIENCY UNQUOTE END MESSAGE.

Staggering news. The D.S.O. was the next most important decoration after the Victoria Cross. There was no false modesty in his surprise: he reckoned he'd been going pretty well lately, but he didn't think his work had been *that* outstanding! He was very pleased about the wording of the official announcement, because it was a tribute not just to him but to the whole squadron.

It didn't strike him that there was a fine justice in the fact that his D.S.O. had come through in the final hours of 1940—the grimmest, greatest year Britain had known in centuries. A year which had given him his first opportunities, pitched him into battle, tested him to his limits, aged him—and now finally rewarded him with high honors.

January descended upon them—adamant, with its gray marble skies, slithering fogs and ice-barbed gales, making the airmen of both sides its prisoners, cooped up in their fuggy dispersal huts for days on end. Even the resolute Luftwaffe night-bombers stayed on the ground most nights, and Britain enjoyed a period of comparative peace. Somehow Tuck didn't resent the winter as much as in past years.

So that she could stay near Coltishall, Joyce quit her war job with the London Ambulance Service. Tuck got her a "hush-hush" post as chauffeur to the local branch of the Ministry of Aircraft Production. Now they could be together at the weekends and two or three evenings during the week.

On Sundays they sometimes went for a drive or hired horses from a nearby stable and rode deep into the country. The evenings were spent drinking or dancing with some of the other Coltishall pilots and their girl friends, or playing records in the lounge at home. Fortunately Joyce's musical tastes veered towards the classics and she soon came to share his love for the great symphonies.

Strange how Tuck—so restless in the mess, and quite unable to relax even in his own family's home—now managed to sit quietly by the fire for hours on end. Sometimes they'd be joined by Joyce's mother, Maude, and her stepfather, "Nunkie." Maude, a stately yet bustling woman, would busy herself with letter writing, embroidery or knitting, the men would talk and Joyce would lie back in her armchair, listening, with her eyes closed most of the time.

Nunkie—Mr. Ackermann—was a small, benevolent man with a brilliant, crinkly smile. He'd made a fortune out of an art and antique gallery in Bond Street, London, and retired many years before at the remarkably early age of forty. He read a lot, still collected antique furniture, paintings and silver, enjoyed long, leisurely country walks in almost any kind of weather, liked to go shooting or fishing occasionally on the Norfolk Broads. He also took great interest in local affairs, and organized and supported a number of charity projects.

All told, those evenings at Joyce's home that winter were a great boon to him. The house was a balm that soothed away much of his nervous fidgetiness and provided fresh interests for his care-burdened mind. As a result he became much more tolerant and reasonable. Until he met Joyce, he had been well on the road to becoming passionately inflexible.

From time to time one or two of the North Weald pilots would drop in at Coltishall for a natter and a drink. Tuck was disturbed to hear that Victor Beamish was still "riding herd" on his wing. There was, of course, absolutely no need for this now, but having started flying regularly again Victor simply refused to give it up.

One evening when Joyce was on late duty Tuck took off in heavy rain and groped his way through to North Weald. Victor was delighted to see him, and they drove some distance from the airfield to a quiet little pub where they were unlikely to run into any other pilots. After a couple of hours they were both rocking gently on their heels and slurring their words a bit. Suddenly Tuck slammed his mug down on the wood and said: "Victor . . . I wish to God you'd pack in this lone-wolf stuff!"

Beamish regarded him thoughtfully for a moment, an uncertain smile tugging at the corners of his mouth and narrowing his eyes.

"Now then, Robert. . . . So this is whoi you've com' all this way, is it?"

Tuck hesitated, then shrugged and nodded admission. "You can't do it, Victor. You *know* you can't. Nobody can get away with it indefinitely. And you've had a helluva long run."

"Sure and you're just the man to be lecturin' me, eh? You, with your careful ways. . . !" He sucked on his stubby pipe, making that sputtering sound, and a dribble of saliva ran down to his chin.

Tuck reached out quickly, gripped the Irishman's forearm and stared into his face as if trying to hypnotize him.

"Victor—sir!—*don't do it!* They'll get you, sure as eggs. For Christ's sake, d'you think they're blind? D'you suppose they haven't noticed you every time, 'way out behind on your own? Make no mistake, they bloody well know all about you!

"Shall I tell you what'll happen?—I don't *need* to tell you, do I?—because you know damned well yourself! They'll set a trap for you one fine day. They'll be waiting up-sun and they'll let the whole bloody wing go by, then they'll all drop on you and——" He made a short, violent, squirty sound with his lips compressed. "Don't do it, Victor. Don't do it any more."

Beamish stared down into his mug, moving it in a small circle, making the beer swirl and slop and foam. Very quietly he said: "Oi'm not a blitherin' idiot, y'know, not entoirly, and what you say may be just about the soize of it. But Oi'll look after me own problems, thank you. So do me a kindness, Robert, and bloody well belt up."

Tuck sighed, shook his head, then grinned wearily. "Oh, balls!" he said, and ordered another round.

★   ★   ★

He received his D.S.O., and the previously awarded bar to his D.F.C., from the King in an investiture ceremony held at Bircham Newton, Norfolk, a big Coastal Command base, on the afternoon of January 28th, 1941. (The date is another of those astounding coincidences that color his life: exactly one year later *to the very hour,* he was shot down by the German batteries at Boulogne.)

Joyce could have had the day off and watched from a spectators' enclosure, but he wouldn't hear of that. Never since he'd left home to join the S.S. *Marconi* had he allowed any one of his friends or family to see him off at a railway station or attend any kind of service ceremony or display in which he was involved. His only companion on the big day was a cheery young flight commander named Van Mentz, from 222 Squadron—also based at Coltishall—who was to receive the D.F.C.

It was a very impressive ceremony, with bands blaring and neat boxes of white-belted airmen on parade in front of the giant hangars. It had snowed the day before, and it was still bitterly cold, with low, gun-metal clouds, and little islands of hard-packed snow squeaking under the marchers' boots.

Throughout the ceremony three twin-engined Bolingbroke aircraft patrolled round and round the base at a few hundred feet, wings rocking in the turbulent air—keeping vigil just in case a lone Hun tried for a hit-and-run raid.

The King was accompanied by the Queen and the two young Princesses. As Tuck stepped forward he noticed that Margaret, seated on the platform behind her father, was intent on watching the circling Bolingbrokes.

"I am glad to see you again," His Majesty said. He spoke very slowly and deliberately, as if mentally polishing each word before uttering it. He looked drawn and tired, but not in the least nervous: his handshake was firm and his smile confident. "I expect you are having an easier time of it now, with this weather?"

"Yes, sir, things are a lot quieter."

"Ah well, it will do you good to relax for a change. By the way, where was it that I decorated you before?"

"Hornchurch, sir."

"Yes, I remember now." He smiled suddenly and glanced at the sky. "It was a rather better day than this." Tuck smiled back, and the King presented him with his medals. "Once again, my congratulations."

At this moment he suddenly thought of the Iron

Cross, still in his left breast pocket, separated from the British decorations by a single layer of fine fabric; he wondered what His Majesty would have thought had he known. . . .

"Thank you very much, sir." As he took the regulation two paces backwards and saluted, out of the corner of his eye he saw the younger Princess still craning her neck to watch the planes patrolling above.

When the ceremony was over and the Royal visitors were about to get into their cars, ten-year-old Margaret suddenly tweaked Tuck's sleeve, pointed upwards and said: "Tell me, what kind of airplanes are those?"

Everyone stopped and turned toward them.

"These are called Bolingbrokes, Your Royal Highness."

She frowned and cocked her head to one side. "Aren't they Blenheims?"

"Well—yes, as a matter of fact a Bolingbroke is just a Blenheim with a long nose built on."

She beamed, triumphant, and yelped: "I jolly well thought so. Pa-pa said they were *Hudsons!*"

His Majesty was the first to roar with laughter. Even Elizabeth momentarily forgot her earnest formality and, chuckling and shaking her head, took her little sister's hand and led her on to the cars.

*   *   *

On the day that 257 were given four-cannon Hurris, a jubilant Tuck began to lay plans for an offensive sortie over Belgium or Holland. According to Air Ministry, neither the Hurricane nor the Spitfire could stay in the air long enough, or travel far enough into enemy territory, to make such operations worthwhile. He was determined to prove otherwise—he remembered very clearly the confusion he'd managed to sow back in June, when Daddy Bouchier had let him take Allan Wright and Bobbie Holland on a brief sweep over France, and he knew Sholto Douglas favored a swing to the offensive.

With Chief Tyrer and Cowboy, he worked out a system by which they could log the fuel consumption, range and endurance of the Hurri to the gill, yard and second. From then on he and Cowboy snatched every chance to get into the air—even if only for half an hour or so before the weather closed in again. At low altitude, they flew in formation up and down a measured stretch of coast, experimenting with different throttle settings, engine revs, fuel mixtures and speeds, keeping a meticulous record on each flight. After about three weeks they had amassed a pile of data which proved that, given at all reasonable wind conditions, the British fighters could make a round trip as far as the north of Holland, and during their spell over there have enough spare fuel to engage the enemy for ten minutes using full combat power.

Tuck prepared a report, attached the log sheets and sent the lot to Group. Two weeks passed without news, so he asked permission to attack ground installations across the Channel. This request was refused. Disgusted, he gave up the whole idea for the time being.

The weather began to improve, and enemy activity increased proportionately.

Late on the night of February 10th three Me. 110s raided Coltishall. Tuck was down at dispersal, and he heard their screaming engines as they dived in to attack. He ran out of his office just in time to see one of them flash across a patch of moonlight and latch on behind a 222 Squadron aircraft which was about to land after night patrol. He was quite unable to stand and watch—he found himself running out over the grass yelling "Look out!" at the top of his lungs. But even if the 222 pilot could have heard the warning it would have made no difference, for the very next instant the raider opened up and blew him out of the sky, a flaming wreck.

Tuck sprinted to one of the defense posts, jumped into a slit trench and began blazing away with a twin Lewis gun. If there was one thing he hated, it was being caught on the ground like this. . . .

Geysers of flame and dirt rose all around and the air shuddered. The wicked whistle of bomb splinters lacerated the night. The Germans' cannonshells and bullets flared across the grass and the perimeter track like quick burning fuses. There was a stench of cordite and a sort of deep, steady trembling in the ground. Other Lewis guns were going now, and somewhere a fireball clanged hysterically.

He reloaded and kept firing at the shadowy shapes swooping low over the field. The bastards weren't going to destroy *his* parked aircraft! They must not smash his Hurris! He kept shouting to this effect all the time, like a maniac.

It was all over in two minutes. There were some big holes in the field, several barrack blocks and other buildings were severely damaged, and one or two airmen were wounded. But none of the 257 machines was damaged. Tuck had yelled so much he was hoarse as a crow.

"By god, why can't we give them some of their own bloody medicine?" he rasped to the world in general as he stomped into the mess an hour later. "When are they going to let us go over there and shoot the hell out of *their* fields?" After he'd washed the dust out of his throat he went to his room and drafted a signal to Group, asking once more for permission to organize an offensive sortie.

But again he was turned down.

One of their favorite drinking haunts now was the *Ferry Inn* near Horning, which stood quite on its own in open country. Pilots from all the Coltishall squadrons were in there three or four evenings a week, and one Saturday night in the saloon bar Cowboy was introduced to a smokey-eyed redhead named Jonnie. Somehow he pried her away from her friends, and she told him she was a widow—her husband, an army officer, had been killed in France. He dated her, and for the next two or three weeks took her out every night he wasn't on duty.

At Cowboy's invitation Joyce and Bob joined forces with them once or twice, but neither of them liked Jonnie much, though they weren't sure why. They found excuses to stay on their own.

One night when Joyce was on a late job, Tuck remained in the mess until well after midnight. He was just leaving to go to bed when a steward said: "Telephone sir—a lady." He hurried into the anteroom to take the call: it didn't dawn on him that it could be anyone but Joyce.

"Hello, darling," he began—but a strange voice answered, high and shaky.

"Bob . . . thank God I've got you! This is Jonnie. Listen, you've got to come at once. Don't ask questions—there isn't time. Just come quickly!"

"Hang on, old girl . . . d'you know what time it is?"

"Please, Bob! Take my address—write it down. For God's sake do what I say! *Cowboy's in a helluva spot!"*

That did it. "All right, fire away!" He fumbled for a pencil and scribbled the address—a flat in Coltishall—on the back of a cigarette packet. "Roger—there in five minutes."

He made it in four. Jonnie was waiting on the outside steps, trembling and tear-streaked. She was wearing a light silk dance dress, and she seemed to have been drinking a bit. He couldn't make much sense out of her babbling, but one phrase registered all right—"... my husband ... home on leave, didn't expect him . . ."

"Where are they?"

"Top flat."

He went up the narrow stairway two at a time. The door of the flat was open. He heard a slurry male voice —not Cowboy's. He paused for a second on the landing and listened, but couldn't catch the words. Gently he pushed the door open and tiptoed in.

From the little hallway he could see into what seemed

to be the sitting room. Cowboy was sprawled on the sofa, holding a tumblerful of Scotch with both hands. He had his tunic off, his tie was loosened and his hair was mussed. He was wearing a silly grin.

Standing in the center of the room, swaying like a sapling in a nor'-easter and murmuring incoherently, was a small, plump figure in the uniform of an army captain. He held a heavy Colt revolver—far from steady, but pointed in the general direction of the Canadian's chest.

Even as Tuck took all this in and tried to decide what to do, Cowboy suddenly threw back his head and released a bellow of laughter.

"Jeez, you're not only drunk, pal—you're crazy! But I'm gonna tell you one thing, pal—you haven't the guts of a louse. If you had, you'd have pulled that trigger long ago, 'stead of standing there spoutin' like a goddam schoolma'am."

The little captain blinked, stiffened and struggled to speak with clarity.

"Dirty bas-tard . . . show you who's scared . . ." He lurched a step closer. Cowboy laughed scornfully again. Tuck started into the room, slowly and quietly, moving round to get behind the captain. But Cowboy spotted him at once and called, *cheerfully:* "Hey, here's Bobbie! Hiya, chum! Give the man a drink, Pongo!"

The captain wheeled round and staggered against a coffee table, upsetting a stack of empty glasses. Then he backed quickly to the far side of the room where he could cover both of them. Tuck saw that Cowboy was right: the man was frightened—*too* frightened.

He decided there was nothing for it but to try to use the authority of his rank.

"Captain," he said in a loud, barky voice, "what the hell d'you think you're doing with that gun?" The man wagged the revolver at him.

"Gerrout . . . Gerrout, damn you."

Tuck drew himself up, glowering, affecting Blimpish wrath.

"Don't talk to me like that!" he rapped. "Put that gun down at once, you fool, or you'll find yourself in serious trouble. I've come to take this officer back to his unit—I'm his commanding officer—and I won't stand any bloody interference from you or anybody else!"

For a mere second or two the captain hesitated and Tuck risked a step forward. But at this point Cowboy almost wrecked everything by laughing again. Viciously Tuck turned on the Canadian and yelled: "Up off there, you useless sod! On your feet when I'm talking to you!"

Cowboy's laughter died. He gaped at his friend, and saw his nostrils pinched with anger. He got to his feet very slowly, as though unfolding joint by joint.

"Get your tunic on," said Tuck, then he turned back to the captain. The little man stared with large, glazed eyes, and his mouth opened and closed, but he made no sound. Tuck snatched his chance, strode directly to him and held out his left hand.

"Give me that gun."

There was a long, breathless pause with the glazed eyes narrowed slightly and flicking uncertainly from Tuck to Cowboy. Then the captain seemed to wilt. His head slumped forward, he leaned back against the wall and allowed the Colt to be taken from his drooping hand.

Still carrying the gun, Tuck grabbed Cowboy's arm and bundled him out of the place. At the foot of the steps Jonnie waited, very quiet now, shivering a little.

"That man *is* your husband?" Tuck asked. She didn't look at Cowboy.

"Yes . . . but I never expected to see him again. We quarreled last time. . . . He said he was going for good, and he never wrote."

He was afraid she'd start crying again. He shoved Cowboy out of the door—"Wait in the car." When the Canadian had negotiated the steps and clambered into the Hillman, the girl caught Tuck's arm. "What am I going to do? I can't go back up there. . . ."

Very deliberately he pulled her fingers from his arm and drew away. He had no glimmer of sympathy for her. She looked into his face for a moment, then turned and walked slowly to the foot of the stairs. She leaned her head on the big, square banister post, and her shoulders shook under the silk dance frock. Tuck went out and shut the door very firmly.

Back at the airfield he went into the kitchens and rustled up a mug of black coffee. Cowboy drank it in short, shuddering sips, and sobered up a lot.

He told Tuck that Jonnie had thrown a bottle party at the flat, and about twenty people had turned up. In the middle of the evening the little captain walked in. At the sight of him Jonnie went gray as old death.

"Didn't you know she had a husband?"

"Naw, I'd fallen for that merry widow line. I still believed it. I sure liked that girl, y'know, Bobbie. . . ."

Bit by bit the rest of the story came out. Jonnie had taken the captain into another room, and when she came back she was still very pale, but when Cowboy asked what was wrong she said she'd explain later. The captain returned and did some very earnest drinking, sitting by himself in a corner. When the party broke up around half-past eleven, only the three of them were left in the place—and then came the showdown.

The captain, now nasty drunk, revealed his identity and bawled abuse. Cowboy was too plastered himself to understand the seriousness of the situation. Suddenly the Pongo produced the gun and Jonnie darted out of the flat, terrified.

"He didn't go after her—he was too shaky on his pins. He just made me pour two whiskies and said we were going to have a last drink together, and then he'd shoot my head off! That's about when you came in, Bobbie." He straightened in his chair, considering his own words, and all at once his brow pleated and he gave a little shiver. "Christ, I guess that really was a close one. The little sod—he might have bumped you off, too!"

"Pete—stay away from that woman."

Cowboy held his gaze for a moment then dropped his head and said quietly: "Okay, okay. But what d'you reckon will happen to her?"

"I hope Soldier Boy tans her arse. Oh, don't worry,

it won't be worse than that—I've got the gun in the car. Going to keep it, too. I hope they bloody well court-martial him for losing it."

In a couple of days they learned that the captain had returned to his unit in the North of Scotland, and taken Jonnie with him.

*   *   *

A few nights later, Tuck and several of his pilots were bounced by hit-and-run raiders in pitch darkness as they entered the circuit to land after patrol. In the first moments of confusion, as control suddenly shrilled "Bandits in the circuit!" Sergeant Truman's machine was blasted from close astern and set on fire.

"Climb and bail out!" Tuck ordered him. "Don't try to land that thing. D'you hear me?" There was no reply. He kept repeating the order. "Truman—do as I say. Bail out. If you attempt to land that aircraft I'll have you court martialed. Climb her and bail out!" He was so intent on the sergeant's predicament that he wasn't taking evasive action or keeping any sort of look-out, and risked being clobbered himself. Truman's radio may have been smashed, or the lad may have been too shaken or badly hurt to respond. He continued to make his landing approach. As he neared the boundary the burning Hurri dropped its nose and flew straight down. A tower of yellow flame rose in the night.

The intruders dropped a few sticks of bombs and made off. The Hurris had no fuel left and couldn't give chase.

Truman had been married quite recently, and his wife was staying at *The Recruiting Sergeant* on the main Norwich Road. Tuck had met her there once or twice. She was a pretty little thing, frighteningly young. For some reason never to be known, Tuck decided he must break the news to her himself.

He drove over to the pub, frantically composing phrases in his mind. Nothing seemed even half right. He was no good at this sort of thing—he should have left it to the adjutant, or the chaplain.

He woke up the landlord and his wife and said simply: "I must see Mrs. Truman right away." They showed him into the little parlor bar and after a few minutes he heard the girl's slippered feet scampering down the stairs. He groped for the phrases he'd worked out, but he'd forgotten them all.

She raced into the room and stopped short in front of him, a slight, large-eyed girl, still half-child, clutching a wooly dressing-gown over her bosom. They looked at each other, and in the lamplight he distinctly saw the blood drain from her cheeks. He knew there was no need for words. She had read it all in his face.

Suddenly she swayed forward and her head smote his shoulder, hard. Her face was pressed against his wings brevet, and he could see the tear streams soaking into the cloth. The small, slender body shook as though great blows were being rained upon it. He put his arms round her and held her there, firmly, for what seemed a very long time. Then the landlady gently pulled her away and led her back to her room.

He had spoken not one word.

The landlord came in and handed him a glass of rum. He drank it gratefully, and told him: "The adjutant will be over in the morning. Things to be sorted out, you understand."

"I understand," the publican said. "When you run a pub near an airfield these days, and let off your rooms to wives and sweethearts, you soon learn what to expect when there comes a knocking on your door in the middle of the night."

The next to go was the gallant Carl Capon—shot down in unknown circumstances while flying alone on another night patrol. As Tuck's number-two ever since the great September battles, Carl had served faithfully, skillfully, and with great heart.

Capon—the watchful old tick bird, picking the Messerschmitts off his back. . . .

Tuck would never let any of the others see his grief —only Joyce sensed how much he missed this loyal and reliable "shadow." The loss set him thinking again about Caesar, and for the first time he told her about his old friend. Strange how he talked of him— hesitantly, half-penitent, almost as if he were telling her of a past love. She listened, but as usual said very little. And in matters of such a personal nature, that was the way he liked her to be.

To replace Carl he chose one of the new boys, Flight Sergeant Ronnie Jarvis, a wartime volunteer, highly intelligent and well-spoken. Before joining up, Ronnie had been one of the most promising young executives employed by the great financier Isaac Woolfson.

One afternoon, with Ronnie's support he intercepted a Dornier 17Z and after chasing it far out to sea destroyed it from almost maximum range. Praise the Lord for 20 mm. cannon!—machine guns could never have done it. . . .

The fast "Z" had been difficult to overhaul, but after the first flights of shells had smashed into the wings and fuselage it slowed and started down towards the water leaving that familiar, slim black trail—like a giant spider spinning its thread. Then they were able to close the range quickly, though there seemed no need to clobber it again. As they drew near, the rear-gunner, who by this time should have been thinking about bailing out or bracing himself for the impact of ditching, opened a defiant and coolly accurate fire. Before Tuck could swerve out of the way, bullets shattered the left sidepane of his windscreen and lopped off his rear vision mirror. He broke off and climbed to one side, with Ronnie still guarding his tail. Still the German stuck to his guns, reaching for the attackers with his lashing tracer, almost until the moment the bomber hit the sea. They went down and searched for several minutes, but spotted nothing but oil slicks and a few fragments of one wing.

Tuck had been impressed by the German gunner's stubborn action. A courageous soldier and a fine marksman had gone down, like so many other admirable men, into the black, bottomless ooze. Somehow it seemed a totally unfitting end for men who'd roved and fought in the clean, clear sky.

Wonderful news awaited him when he landed at base. Group had signaled permission for an offensive sortie over the Continent the very next day. He was to lead the first strike against the enemy-occupied territories to be made by fighters from East Anglia.

14

Group's briefing imposed unexpected restrictions: only two Hurricanes were to take part in this "experimental offensive," and in no circumstances were they to open fire unless attacked by enemy aircraft. When he read these orders Cowboy expressed his rancor in drawled, listless oaths, but Tuck was surprisingly philosophical.

"Never mind, it's a start—thin end of the wedge," he said. "One thing's certain, Sholto Douglas is flat-out for sending us across as often as possible, and if we pull this one off without trouble you can bet we'll strengthen his arguments and he'll have his way. Now stop binding, Pete—or by God I'll take someone else along. . . ."

Their objective was a good-sized, triangular area of Holland which included several Luftwaffe bases, some important railway junctions and bridges, and one or two factories. Their main purpose, the orders emphasized, was to test the strength and efficiency of the defenses —the Germans were believed by now to have lined the entire length of their "Atlantic Wall" with A.A. batteries. Conscientiously, the two pilots spent the evening poring over maps of the region and working out their flight plan, double-checking every detail.

Next morning the clouds connived at their plot. An almost unbroken layer of soggy stratus stretched northeast, drifting sluggishly at only a few hundred feet. The meteorological experts said it extended right over Holland. This would provide excellent cover if they needed to make a quick climb away, and yet provided they stayed fairly close together it wasn't thick enough inside to hide them from each other. They planned to maintain complete r/t silence until they'd crossed the Dutch coast.

Quite a crowd came down to dispersal to see them off. It was a long, long time since any British fighter had sailed boldly into the enemy's stronghold. This was an exciting moment, a minor turning point which gave every pilot the right to hope for some fine hunting east of the Channel in the spring and summer days to come —and ah, that was a *sweet* thought after the grim struggle of the last eight months!

But in the very first minutes after take-off, Tuck thought he'd have to cancel the whole thing. As he throttled back to the economical cruising speed computed in their flight plan, the Hurri began to wallow unpleasantly, very sloppy on controls and threatening to stall. He checked his engine settings—r.p.m., mixture, pitch. All quite correct, and the engine sounded smooth and strong. What the hell could it be, then?

He glanced over at Cowboy and saw his chubby face riven by a huge grin. The Canadian held up one hand, with two of the fingers extended downwards, and he wiggled the fingers as a child does to imitate walking legs. Then he nodded down under Tuck's wings.

The undercart—in his excitement, for the first time in his entire career he'd forgotten to retract the wheels! His face burning, he snatched at the lever and jerked it to the "up" position. The heavy wheels and legs which had been dragging in the airflow folded into their compartments and the aircraft was instantly bouyant and manageable.

He was furious with himself. He stared rigidly ahead for several minutes before he turned and gave Cowboy a sorry grin.

The two Hurricanes bored on through the gray day, and a little way out over the North Sea they dropped to low altitude, skimming the heavy green rollers to avoid having their pictures taken by radar screens on the Dutch coast. It was a long haul, and there was plenty of time to wonder what it would feel like trying to nurse a shotup engine back across this expanse of ocean. It was near-freezing down there, and the rescue boats would have a tough time trying to find a tiny dinghy this far out in such poor visibility. Death by exposure must be very unpleasant. Tuck wished he could talk to Cowboy on the r/t—that would have helped a lot. . . .

But after all this mounting tension, the fifteen-minute period over Dutch territory was a huge anticlimax!

They saw plenty of fighters, all on the ground. Plenty of Germans, too, but the majority took only casual notice of the lowflying Hurris. Around the airfields and dotted along the coast were many gun positions, but only a few of the crews spotted the British markings and came to life in time to fire wild bursts after them. Some of the pot-helmeted gunners, lounging by their pieces, actually waved cheerfully—taking them for a couple of Luftwaffe boys out on a spree. Not that the Hurri resembled the Messerschmitt—apparently it just didn't cross the Germans' minds that British fighters could come this far. . . .

They circled one large airfield several times, then flew straight at the control tower at only five feet above the grass. As they pulled up and banked steeply, grazing the top, they glimpsed startled upturned faces behind the wide observation windows.

Next they roared close over the heads of a small party of troops trudging along a main highway, forcing them to break ranks and duck; the officer at the head of the column waved his fist at them, jumping up and down with rage—yet no rifles were raised.

But the Dutch people were different—*they* recognized them quickly enough! When the Hurris wheeled round one large farmhouse, the entire family seemed to come rushing into the courtyard, waving towels and aprons with unmistakable joy. Workers in the fields and people in village streets doffed their hats and whirled them at arm's length. It was all very heartening. Tuck was thrilled.

Then, keeping an eye on the time and the fuel gauge, he started working round west and soon picked up the main road which would lead them to their "exit" point. On the way they came upon another, larger column of marching troops. There were trees on either side of the road, so this time they couldn't dive very low. More cheerful waves, upturned smiling faces. . . . If they knew how his thumb itched for the firing button!

We're coming back, Fritz, we're coming back! Better wake up, you dozing bastards—stir yourselves and start watching the sky—all day, every day . . . because from this moment on you're never going to be sure where or when we'll be dropping in! This far you've had it all your own bloody way, but now our innings is starting. So get your fingers out, you stupid sods, and dig yourselves some slit trenches. . . .

Fine, proud thoughts—a trifle histrionic, maybe, but wonderfully refreshing! He looked over at Cowboy and as their eyes met, across forty feet of rushing air he could feel his sympathy and controlled excitement as poignantly as if they had been talking together in a quiet corner of the mess.

They crossed the coast unmolested, and as they neared their home shores they were so comfortably within their fuel "budget" that when they spotted a lone Me. 110 away to the north, they were able to turn after it, using full throttle for perhaps four minutes— until it spotted them and scuttled into the overcast. They headed for base, and landed with their gunport seals intact.

This was the only occasion on which Tuck got within range of the enemy and withheld fire. He never refused combat—it can never be said of him, as it is sometimes said of other aces, "He only gets away with it because he chooses the moment to fight and the moment *not* to fight. He often backs down when conditions aren't in his favor." It would be understandable if he declined to talk about this one time when he had to watch lush targets pass through his sight and yet, by order of his superiors, keep his guns mute. But in fact he recalls the operation with pride—and remembers it in considerably more detail than some others in which he got "kills" or was shot up himself!

The profound impression it made on him was justified. Subsequent events showed it had been an historic flight. It set the pattern for the type of operation which

Sholto Douglas christened "Rhubarb"—for in addition to his plans for large-scale fighter sweeps over France and the Low Countries in the bright days of the coming spring, the C.-in-C. envisaged countless sorties by single fighters and pairs on days when the clouds clamped down and big operations were ruled out. Tuck's report proved that the "Atlantic Wall" had big gaps, that the German radar could be evaded, and the Luftwaffe caught with their tails down just like anybody else! It was exactly the sort of ammunition Sholto Douglas needed to subdue the last of his critics at Air Ministry.

*   *   *

Pete Brothers was finally "discovered" by Group— as Tuck had long feared he would be—and posted away to command a new Australian squadron. He was replaced as flight commander by a stumpy, fair-headed officer with a swooping cavalry mustache—Peter Prosser Hanks. Prosser, as he was invariably known, had distinguished himself in the early days with the famous No. 1 Squadron in France, and was one of the most brilliant pilots in the service. Tuck remembered meeting him at Tangmere long before the war.

Orders came through for all Coltishall squadrons to practice low flying in formations of twos and threes. They obeyed with high enthusiasm, and even the civilians over whose rooftops they roared were delighted— because they too realized the significance of this new activity.

About this time the lone night raiders began to concentrate their attentions on Lincolnshire, where a large number of training airfields now thrived. The intruders would slink inland, look for a field where night-flying training was in progress, then join the circuit and in the space of a few minutes create appalling carnage. The unarmed training kites—Tiger Moths, Harvards, Avro Ansons and Airspeed Oxfords—burst into flames very easily under the cannon and machine-gun fire, and for every one that was shot down two or three others, manned by wobbly weanlings probably on their first night solos, fell out of control and crashed. By

these tactics, the Luftwaffe doubtless hoped to whittle
down the supply of new pilots to R.A.F. operational
squadrons.

To help the night-fighters and ground defenses guard
the training fields, several day squadrons were brought
in to patrol up and down the east coast during darkness.
No. 257 drew a "night beat" which extended from
Skegness southwards to the Thames Estuary. They operated singly, directed by ground control but without
any sort of equipment. Sometimes they were in the air
until two and three in the morning. It was monotonous,
tiring work after a normal day's flying. But Prosser
soon proved it all worthwhile by bagging a Heinkel,
so they stuck to their task in good spirits.

On March 30th Tuck was awarded a second bar
to his D.F.C. The official announcement said, "for
conspicuous gallantry in attacking enemy raiders."

Three D.C.F.s: in the entire history of the Royal Air
Force, only one other pilot had achieved this "hat
trick"—Flight Lieutenant A. L. Turner had been
awarded a second bar just one month before. Again
he was quite genuinely taken by surprise, and at the
same time delighted.

He didn't know what to say when, one by one, his
pilots congratulated him. They expected him to make a
little speech in the bar when they proposed his health,
but he pretended not to understand that—he just
laughed and said "Cheers," feeling like an actor come
to his first night without having learned his lines. It
developed into another big beer party, of course, cul-
minating at the *Ferry Inn*—and this time Joyce did
join in.

For a while Joyce saw little of him, because most
evenings now the squadron was either held at readiness, or up searching the inkwell heights for the enemy
cloud-sappers who tried to burrow their way through
to the heart of Lincolnshire. Tuck worked unsparingly. Day after day, night after night, the fat-tired wheels
cut their fast tracks in the soft, wet grass, and the
big, tough, slugging Hurris seemed permanently to hov-
er, clamant, above the Norfolk fields and townships.

On the night of April 9th Tuck and Prosser, patrolling several miles apart, were both directed after a Ju. 88 coming in towards the coastal town of Lowestoft. The raider was tunneling westwards through a thin, smooth layer of stratus around 2,000 feet which roofed in the whole of East Anglia and some of the North Sea, shutting out the moonlight. But it would have to drop down out of the cloud as it approached the coast to try to check its bearings before making its lunge inland—and that was the moment when the hunters would have their chance.

Under the stratus it was very dark. Tuck, leaning well forward in his seat and staring hard, could distinguish nothing beyond the windscreen. He was confronted by flat, adamant darkness—the outside of the perspex might have been coated with tar.

Control gave him a small course correction and added: "Bandit descending now . . . descending . . . leveling out." A pause, then: "Angels one thousand!" He acknowledged, put the nose down and descended in a shallow dive. Then, straight ahead, the night was ripped open by a series of blinding flashes. A.A. barrage—from the ground they could probably just make out the bomber's silhouette against the light gray canopy of stratus. A golden web of tracer was filling the sky in front. He'd have to turn off pretty quickly if he didn't want to run smack into the stuff!

He veered round to starboard. More flashes, a bit closer this time. And in the last one—in that tiny fraction of a second when the exploding shell split the darkness like a bolt of lightning—a dim shape was printed on his eyes. The shell flash had died and the darkness had slammed in his face again long before he actually saw that shape—or rather, saw its image, the slowly fading memory-picture held by his light-shocked retinas. He seemed to study it almost at his leisure before reaching a number of conclusions—*Ju. 88, range uncertain but not more than six hundred, course roughly north-west, crossing almost directly ahead and a little higher.*

If only he could get round quickly and maneuver directly behind it, there was a faint chance of picking up the glow of its exhausts. Holding his height steady, he shoved the throttle forward, stood the Hurricane on her port wingtip and whipped her tightly through sixty degrees on to a nor'westerly course. He leveled out quickly and then used his rudder to skid her gently from side to side.

Luckily the A.A. barrage now stopped, as suddenly as it had begun. Control must have warned them off, to give him his chance. He hunched his shoulders and thrust his jaw forward, and in the pale green radiance of the instrument panel his lean face took on a wolfish look.

Left rudder . . . right rudder . . . left . . .

The fighter maintained a steady course and her wings stayed level, but she slid crablike first one way, then the other, progressing across the night like a skater on black ice. After about a minute-and-a-half, he began to increase the amount of rudder, skidding a little wider each time. At the end of one of these movements he caught a tiny, dull red flicker above and to the right—a mere pinpoint, like a distant star glowing through mist. He seized on it, and hung on with his straining eyes, willing it not to fade—afraid he might have to blink, and that it would vanish in that instant! Holding his breath, he wrenched his aircraft round and climbed towards it.

Slowly the flicker grew and brightened, until it became a sputtering little puff of flame. Aircraft exhaust. But—Prosser was somewhere near and perhaps other Hurris too; only if he saw *two* exhausts, very close together, could he be certain that it was the twin-engined Junkers he'd latched on to. He throttled back a bit, so as not to overtake too quickly, and then very, very gently he began to skid from side to side again.

Delicate work, this. A mite too far out from astern of his quarry, and he'd lose sight of that sputtering flame—and most likely never pick it up again. It was like walking a tightrope, blindfold.

A full two minutes of this, with the sweat breaking out all over him and fast growing icy in the chill night air, then suddenly there were two exhausts staring in at him like mad, red eyes. And between them: the faint glint of metal. It was the Junkers, all right.

The rest was easy. Dropping slightly, he closed to 200 yards, checked the turn-and-bank needle and fired a very long burst into the belly. As the 20 mm. shells smashed into it, the Junkers pitched and rolled like a destroyer riding out a gale. A fountain of sparks from the port engine, cascading in a long curve like a giant firework, then came solid streamers of flame. The Junkers' nose fell and it turned to the right—back out towards the sea. They always turned, instinctively, for home in their dying seconds. . . .

He followed it in the shallow, accelerating dive and dealt it several more short bursts. The whole of its fuselage was glowing dark red now—inside must be an inferno. Like a huge, fiery tailed meteor it passed low over Lowestoft. For a horrible moment he thought it must crash on the rooftops, but it cleared the town and plunged into the sea.

He reported the victory to control in his usual unemotional r/t voice, climbed away and started south, down the coast. His wrist tendons throbbed, his legs were stiff and his back ached: it was as if he had fought a long and violent dogfight.

He noticed that the stratus above him seemed to have thinned a lot. The moonlight was seeping through, making the underside strangely, beautifully, luminous. He sat back on his seat and hummed a snatch of Tschaikovski's Third, filled with a sense of sleepy well-being.

Other 257 pilots had their good nights. Considering that they were without special training, and lacked the "seeing-eye sets" (airborne radar) which were by now standard equipment in Beaufighters and all other night-fighters, their performance was nothing short of amazing. The airwar had changed from the sunlight to the shadow, but the "Burma" boys were still a crack squadron.

One rainy evening he was with Joyce and a few 222 Squadron chaps in the *Ferry Inn* when all at once, quite unaccountably, he grew feverishly restless. He drained his glass and said: "Come on, everybody, let's whip into Norwich!"

Van Mentz—the youngster who'd gone with Tuck to the Bircham Newton investiture in January—looked up at the bar clock and shook his head. "Not worth it. We'd never get there before closing time." The others murmured in agreement.

"That's fast," Tuck protested, clipping the lead on his bloodhound Shuffles. "Pub clocks are always ahead. C'mon—let's get weaving."

A rising wind shook the heavy door, and rain was spanking on the windows. No one stirred. Even Shuffles stubbornly refused to lift his long belly from the warm floorboards in front of the fire.

Joyce watched him, fingering her glass uncertainly, and was astonished to see beads of sweat forming on his brow. She finished her drink and said: "Well, if you're really set on it, we'd better get cracking." He helped her into her coat.

Still none of the others moved. Tuck yanked the reluctant Shuffles to his feet and then made a last appeal. "What about it, chaps? Dammit, we'll make it easily, if you come right away." Robinson, the 222 Adjutant, smiled and said: "No thanks, not tonight." Tuck hesitated a moment more, staring at them; then he shrugged, spun round and strode to the door.

As they got into the car Joyce asked quietly: "Just what are you up to, darling?" He started the engine and got the car moving then said, almost defensively: "What do you mean?"

"Well, this is all a bit sudden! And they're right, you know—we *won't* make it before closing time." She lit two cigarettes and placed one between his lips. He was sitting up, rigid, and driving at alarming speed down the narrow, twisting road. She was sure he was ill—nervous exhaustion, something like that.

"Joyce," he said after another pause, "I just *had* to get out of there."

"Why?"

"Haven't the faintest idea. I suddenly knew I had to go, that's all. Couldn't stay another minute."

He slowed the car and seemed to relax a bit. She put her head on his shoulder. "In that case," she said a little later, "to hell with Norwich. Take me home and I'll cook you bacon and eggs." He didn't answer, but at the next crossroads he turned off for Coltishall.

It was after midnight when he got back to the airfield. Several times during the night Shuffles growled sleepily, and he thought he heard the door open and close softly; once he had the distinct impression of a torch shining on his face. Why should they be checking up on him . . . ? Next morning as he was shaving Cowboy came in, sat on the edge of the bed and said: "Drinking will be the death of you, Bobbie."

"Meaning what?"

"You were at the *Ferry* last night—that so?" Tuck nodded, watching him in the mirror. "Well, for once you must have left before closing time, because just as the bell rang for last orders a bloody Hun came over and scored a direct hit with a five hundred pounder. Bloody grim show."

Tuck turned and stared down at him, his face white as the lather on it. Cowboy went on talking, listing the people killed by the bomb—Van Mentz, Robinson, Attwell the 222 Medical Officer, and six or seven civilians they all knew. The landlord, Albert Stringer, and about half a dozen others had been dug out alive, though most of them were seriously injured. But Tuck heard all this only faintly.

After a while Cowboy stopped short, frowning, rose quickly from the bed and came right up to him.

"Christ, what's the matter with you?" He yelled. "What the hell are you *shaking* for?" Tuck got a hold of himself at once. "I'm all right, Pete," he said gruffly. He turned back to the mirror and went on shaving. His hands were quite steady again.

But just for a moment there, he'd been gripped by a kind of terror. It wasn't just the thought of his narrow escape—but the fact that he'd had some sort of a *warning*. . . . That sudden, unreasonable restlessness —was that what writers called "a premonition of disaster?" The thing was uncanny. It was "Tuck's Luck" carried to an almost supernatural degree.

* * *

Throughout April and May of 1941 the squadron worked hard and happily—night patrols, day patrols, escorting shore-hugging convoys, practicing low-flying and plenty of air-range firing. Very rarely now did they encounter the enemy by day, so for a long time no pilot was lost. Cowboy's "bag" had risen to five, and he was flying with diamond-cut precision. Tuck recommended him for a D.F.C. and the award came through in record time. Another night the Norwich pubs will never forget. . . .

In June the offensive sweeps began, and with great glee they thundered across France and Belgium at "eyebrow level," shooting up railway trains, junctions, bridges, troop convoys and columns, barracks, gun sites, airfields, power stations, radar posts, secondary harbors, shipyards, canal and river barges. They gloried in this legalized vandalism, whooping in delight as locomotives exploded or parked aircraft burst into flame, shouting congratulations to each other like excited football players when a goal has been scored. Tuck relaxed r/t discipline on these sprees, because he knew that by shouting they egged each other on and created a wonderful spirit of gay, reckless devilment—which was exactly what the job needed.

Very seldom did they sight a German fighter in the air during their ten or fifteen destructive minutes within the enemy's realm, and the A.A. fire was laughably inaccurate. Every sweep was an exhilarating adventure. They regarded them almost as little holidays—a reward for the long, monotonous hours of night patrol and convoy escort they'd put in over the last few months.

Inevitably, a number of Frenchmen and Belgians were killed or injured by the strafing Hurris—the pilots knew that, but they never talked about it, because there was no foolproof way of ensuring that only Germans died. Propaganda broadcasts strove to warn civilians to stay away from possible targets, and formation leaders often risked their lives, and those of their pilots, by making "dummy runs" over factories and crowded bridges and then holding off to give people time to run clear before they attacked. In those two or three minutes they took the chance of being bounced by Messerschmitts "scrambled" from some nearby base.

Yet, if a Frenchman was driving a trainload of German troops or ammunition . . . if a Belgian chose to work in a factory that turned out tank parts . . . or if Dutch fishermen swept mines for their conquerors—then the British pilots couldn't be held responsible for what Goebbels termed "the murder of former allies."

They realized, of course, that many people in the occupied countries were forced to work for the Germans, and that some of those "slave laborers" were members of the Underground, engaged on sabotage and passing information through to London. For those brave souls, the chance of being killed by British air action had to be accepted as an occupational hazard. The war had to be carried into Europe, so some Europeans must give their lives in the struggle for liberation.

Occasionally, when the weather was cloudy, one or two of them flew on "Rhubarbs." These were much more sober, wholy premeditated affairs, calling for great concentration and precise navigation. Single, specific objectives were selected in advance: the idea was to get there quickly, hit hard in one blazing swoop, then streak for home—with no dallying, or spur of the moment "side-shows," on the way.

For Tuck and his gang, the sweeps and "Rhubarbs" were too few and far between. But "go easy" was the right policy: Sholto Douglas was stepping up the pace gradually, in proportion to the rise in aircraft production, the supply of new pilots and other vital factors. By the summer everyone hoped they'd be flying regular "milk runs" across the Channel and the North Sea.

L.A.C. Hillman continued to tend the C.O.'s machine with loving care. He claimed the right to stencil each new swastika on the fuselage, and any other airman rash enough to dispute this felt the power of his massive fists.

Hillman usually looked after Shuffles while Tuck was in the air. He would sometimes wash the C.O.'s car without being asked. If Tuck came into a pub where he was drinking, he'd insist on buying him a pint, and press him to a Woodbine.* From the look on the Cockney's honest, battered face on these occasions it was clear that he had difficulty in distinguishing between Tuck and God.

One night when a crowd of them were having a "stomp" in a Norwich Pub, Tuck was warned by the landlord that a constable was about to take the license number of his car—because he'd left it directly opposite a "No Parking" sign, and hadn't bothered to immobilize it in accordance with Defense Regulations. He beckoned Hillman over and said: "There's a copper outside, Hillman boy, and it so happens I don't like the look of him. Just pop out there and knock his helmet off, will you?"

He meant it as a joke—just a silly way of expressing his annoyance, and his contempt for civilian bullshine. The airman was supposed to laugh and say: "Cor, not bloody likely!" or something like that. But not Hillman. He snapped to attention, rapped out: "Yes sir, certainly sir!"—and before anybody could stop him he'd nipped outside.

After a dazed pause, everyone dived for the door. They reached the pavement in time to see Hillman walk up behind the constable—who was stooping to read Tuck's license number in the beam of the his flashlight —and deliver a smart downward blow to the back of his helmet. The helmet was knocked over the officer's eyes, but the chinstrap stopped it coming right off. Hillman, disappointed, was preparing to try again while the constable straightened and staggered back against the car bonnet, but Tuck and the others sprinted over in time to restrain him. They whisked him back into the pub entrance before his victim had managed to get his helmet back in place.

It took a good ten minutes of fast talk to soothe the officer, and persuade him that the whole thing was just an unfortunate misunderstanding. He was a War Reserve constable, a gray and puffy man in his middle

*A brand of inexpensive English cigarettes.

fifties who'd probably joined the force because it was
the best he could do to help his country. It's unlikely
that he believed anything they said, but after a bit his
eyes seemed to dwell on their wings brevets, and on
the medal ribbons that several of them wore, and in the
end he put his whistle away and slammed his notebook
shut. As he started off down the pavement, walking
slowly and sedately, he peered into the pink, cherubic
faces of two of the youngest sergeants and grunted:
"Ought to put you across m'knee, that's what. . . . !"

*   *   *

Joyce enjoyed a good movie. She got him to take
her fairly often, but unless the picture was packed with
action and spectacle he would wriggle and mutter in
boredom. Emotional dramas and light-hearted musicals
he simply couldn't endure.

He was driving her to the cinema one evening when
suddenly, to her infinite surprise, he jammed on the
brakes. Her head almost hit the windscreen. The road
in front was deserted, and she could only conclude that
he'd forgotten something and was trying to decide
whether he ought to return to the airdrome. She started
to speak, but he silenced her with a quick gesture.

And then she heard it—the thrum of aircraft engines, faint, but drawing closer. Tuck threw open the
door, jumped out, peered up into the thick, low cloud.

"Hun—a bomber!" he said, and she saw his jaw
muscles knot hard. There was nothing in sight, but he
knew the distinctive throb of the German engines.

She got out and stood beside him. The sound grew
louder, then out of the overcast came a Heinkel III, so
low they could see its markings clear as hoardings. It
banked over the road about three hundred yards away,
obviously trying to pick up a landmark.

She felt him stiffen in anguish and frustration—how
he hated to stand here helpless and see his natural
enemy so close! If there had been a stone handy, she
believed he'd have thrown it. . . . The bomber leveled its wings and droned off in another direction.
Hardly had it vanished in the grayness when they
heard another plane approaching—a higher, smoother

sound this time. Out of the murk a Hurricane materialized, following the raider's trail exactly.

Tuck gave a high, savage yell and jumped up and
down in excitement. The fighter approached the road
even lower than the Heinkel, and they could see the
pilot's head turning from side to side as he searched
for his prey. Painted on the fuselage were the 257
identification markings and the pilot's letter.

"It's Jerry North!" he screeched. "You're bang on,
Jerry, bang on!" She'd never seen him in such a
state of passion.

But the Hurri crossed the road and carried straight
on—obviously control hadn't yet warned him of the
raider's change of course! Tuck howled in dismay,
and then suddenly he was pelting down the middle of
the road, waving his arms.

"No, no—turn right! *Right,* you bloody clot!" He
roared it out with all his might, as if he really thought
North would hear him. As the fighter continued on and
disappeared, he stopped, let his arms flop to his sides
and cursed solidly for the best part of a minute. Then
he turned and came slouching back to the car.

"This," Joyce said, "is better than the pictures. Let's
just sit here and wait for more little airy-buzzers."
For a moment his face stayed taut and wolfish, then he
gave her a rueful smile.

"Don't be sarcastic," he said.

"I'm not. It was *most* interesting, really, darling.
I've never seen you go off like that before."

He studied her, still smiling.

"You know, sometimes you talk to me like a wife.
And that isn't a bad idea."

A day or two later, Nunkie and Maude somehow
got to know they were thinking about becoming engaged. They made it plain that they were dead against
it—they liked Robert, but they considered his occupation "too precarious." The matter was never openly
discussed, because these views were conveyed by the
most tactful innuendo before Tuck even had a chance
to ask for a private talk with Nunkie.

He couldn't help thinking probably they were right.

Fortune had ridden on his shoulders for a long, long way, but soon the moment must come when it would desert him. He was no good to Joyce—she was entitled to security.

"Let's leave it a while," was all he said to her. "We'll see how things go." And this phenomenally cool woman accepted that without question or a sign of resentment.

* * *

On June 21st, 1941, a fine, warm day, Joyce had the afternoon off. At five minutes past two she was in her room at home, sitting at her dressing table brushing her hair. All at once the big, oval mirror seemed to shimmer and ripple like a sheet of water and she reeled forward, sick, faint, breathless. Inside her a voice was screaming:

Something has happened to Robert!

The sensation passed in a moment or two, but the conviction remained. She wanted to run to the phone and ring up Coltishall, but somehow she resisted the urge. Throughout the afternoon she fought to reason the thought away—mental telepathy was all bunk, like horoscopes and fortune telling! She was overtired, that was all. Her nerves were on edge—she must pull herself together!

She sat on the veranda and read halfway through a novel without a single word registering on her seething mind. As she was going into the lounge at teatime Nunkie switched on the radio, and the newsreader's voice struck at her: "One of our pilots is missing. . . ."

She stood framed in the doorway and as Maude and Nunkie looked up she said quite calmly: "That's Robert. It's Robert that's missing."

And she was right.

He had taken off at 1:30 on a routine patrol, flying alone. It was a wide, bright afternoon and it was good to be back in the dance of the sun. He flew down the coast to Southend without sighting another aircraft.

Without any conscious decision or plan, he put the stick over and headed straight out to sea at 1,000 feet. For sixty, eighty, a hundred miles out he flew, swiveling his head around without pause; all that he saw in the air were a few white feathers of cirrus, high above.

Until the shells came ripping through the cockpit and the engine. And kept coming—a blazing, thundering jet playing on him.

The Hurricane was quivering and rocking and his ears were hurting with the deafening clatters and explosions and his body was frozen rigid waiting for the tearing, bone-splintering blows. Through it all his brain was shrieking hysterically: *fool, fool, fool!* He knew now he'd come out here deliberately looking for trouble. A hundred miles from his coast, no altitude, all alone, perfectly bounced.

This is all the trouble in the world, Robert, and it is all yours.

The hellish din of it—can't possibly last another instant, it must stop now, *now!* Jesus God, how much more is coming in? The next one will hit me, the very next shell—I can feel the place in my back where it'll hit, the flesh is breaking open there already!

God, make it stop: make it stop!

During those few agonizing seconds when the shells were hammering into his machine, panic tried for the stranglehold which would destroy all reason and seal his doom. But Tuck was a seasoned fear fighter. As ever, he hung on, grinding his back teeth, holding back the icy tide that wanted to burst up from the depth of his belly and flood through him—telling himself this torture must end in the very next iota of time. The next, the next, the next. . . .

When at last the din and the shuddering did stop, his mind was instantly clear and his reflexes were working smoothly and swiftly. A 109 flashed by close underneath him. He put the stick forward. The sight dropped on to the flat-topped canopy. He checked the turn-and-bank in the space of a heart beat and squeezed the button. The Messerschmitt waggled angrily under the blast of metal. The canopy crumpled like melting cellophane. It gave itself up to gravity and the deep, dark sea.

The moment he was sure it was going down he

threw the Hurri into a steep turn. Somehow he knew there were others behind him.

Sure enough, as he banked, bright tubes of tracer burned through the air scant feet beyond the perspex. He twisted his neck and saw another 109 very close. It had a yellow spinner: hallmark of a crack squadron. It was traveling too fast to get inside his turn. He let it go by underneath, slammed the stick over and cartwheeled after it. He hadn't been in a real, "split-arse" dogfight for months—he'd soon discover if the old hound could still bite. . . !

The German dived to about a hundred feet, went over on his side and pulled right for all he was worth. Nice flying. Tuck hauled hard and followed round. Round and round. The sea was a flashing vertical wall rushing past his starboard wingtip. The poles spun and the sunlight dimmed. A tremendous pressure on the top of his head and his shoulders rammed him down in his seat. The engine screamed. His eyes ached. The red lines of the gunsight were blurred and bent.

Then the 109 took off a little of his bank. Tuck followed suit and everything came back into sharp focus. The Hun was trapped in the sight. They were still in a steep turn, less than fifty feet off the sea, nevertheless he checked his turn-and-bank needle, and centered it with a tiny adjustment of the rudder before he fired. Two seconds' worth of 20 mm. sent the Messerschmitt straight into the water. Tuck flew through the plume of spray it raised.

He pulled up quickly, aileroned the opposite way, then throttled back. Immediately he was hit from the left. The throttle lever was blown out of his hand. The kick of it numbed every fiber of his left arm, like a high voltage shock. The reflector plate exploded and a chip of it embedded itself in his forehead. The door and the hood flew off, and a tornado raged in the cockpit.

The third German flashed underneath and made a wide circle. The Hurricane's engine was missing badly and he'd no throttle, so he couldn't chase it. He just climbed slowly and watched it come right round and start towards him, head-on. Toe-to-toe stuff now.

The sunlight made the 109's yellow nose shine like a golden bull's eye—an archer's "gold," he thought with mild, irrational amusement. He hoped there was enough ammo left to give the bastard just one good smack on the snout. He was being almost wistful now. In his heart he was pretty sure this really was the end of the road: the Hurricane just couldn't take any more punishment. He'd pushed his luck too far this time—this was the dreaded moment when Fortune would desert him.

He'd no sight, but the turn-and-bank was still working. He wiped a trickle of blood from his left eyebrow and made himself relax.

Charging through the summer sky at a composite speed of around 600, the two fighters slugged it out for perhaps five or six seconds. Marvellously, the German missed. Tuck's shells shattered themselves deep amid the whirring steel parts of the Daimler Benz.

The Messerschmitt made a shaky turn and disappeared eastwards in a shallow dive, bleeding glycol. The German didn't have many miles to go—he'd probably make the coast all right. Tuck knew he'd never make his. . . .

The Hurri's engine was spluttering, crackling, cutting out altogether for seconds at a time. Temperatures and pressures were climbing fast. The radio was dead. The speed was falling steadily: 130, 125, 120. . . . Something back towards the tail was flapping madly. Oil and glycol drenched his legs. The wind tore at his clothes, beat in his face. Blood filled his left eye and ran down the side of his nose into his mouth. His left hand had no feeling in it, and the whole arm was a frigid ache.

Yet, as he coaxed the nose round to the northwest, he was full of deep, warm gladness. Two victories, one damaged. The old hound was still in form!

And still alive!

The wrecked Hurri did her best for him, stumbling on through the shimmering air for an incredible time, until he could actually see a smudge of the southeast coast. Then the starboard aileron fell off and dull roaring flames started coming through around his boots. He had about eight hundred feet. Ah well, over the side it would have to be—that old, familiar torture again. How he detested bailing out! No use prolonging it, though. He undid everything. It was a lot simpler with the door gone: all he had to do was dip his port wing and drop out sideways.

The 'chute opened while he was somersaulting backwards and somehow the shroud lines came shooting up between his legs. His numbed arm regained the power of movement again as he instinctively thrust out with both hands and tried to push the lines away from his face and chest. For a moment he thought he was going to be hopelessly entangled, but there came a mighty jerk, he flipped backwards once more—and then he was descending peacefully and everything on the upper story looked to be in its proper place.

The world was wondrously quiet. He went limp and closed his eyes. It felt like lying in a feather bed. And then *plop!*—he was in the drink.

The water was calm and warm. He shed the 'chute, harness and his flying boots, and got his rubber dinghy inflated. The effort of hauling himself into it exhausted him. He flopped face down in it, head pillowed on his forearms, aching all over but very, very happy. For a while he just lay listening to the sea's soft lappings against the dinghy, feeling the sun on his back, enjoying the smell of wet rubber. Then he lifted his head and with glare-narrowed eyes looked straight into the dial of his wrist watch, just an inch or two from his nose.

It registered eleven minutes past two.

The water must be very salty, it was stinging his hands. He looked at his palms: angry red welts, blisters forming. He remembered grabbing at the 'chute lines. The friction of the thin, silken cords had burned the skin.

Gradually he became aware that an unhealthy amount of water was sloshing about in the dinghy. He sat up and found four small holes in the bottom. Shell splinters. In a little flange pocket he discovered a small "concertina" bailing pump and a selection of tapering wooden screws. He plugged the holes with the screws and quickly bailed the dinghy dry. Everything shipshape again.

He flexed his left arm. The numbness had almost worn off now. He picked at his forehead and to his surprise managed to extract the small chip of perspex. He inspected it carefully, wrapped it in his sodden handkerchief and put it in his pocket. An Iron Cross, a bent penny, a duralumin nut and a fragment of reflector plate—he was building up quite a war-museum!

He clenched his fists and stuck them in his pockets. That lessened the pain of the cordburns. He pronounced himself virtually uninjured and fit for flying duties. Then he lay back with his face to the sun and went to sleep.

The vessel that picked him up two hours later was a grimy, leaky coal barge out of Gravesend. It so happened that the crew had seen a Messerschmitt going down in flames in the distance—very probably one of Tuck's victims. They took him for the German pilot, and as he clambered over the gunwale one jittery deck-hand bruised his ribs with the muzzle of a very ancient sporting gun.

"Now don't *you* start," he said. "I've had enough shooting for one bloody day!" When he'd satisfied them of his identity, the gun was put away and the skipper thrust a half-pint mug in his hand. It was full of rum. When he landed at Brightlingsea that evening he was feeling just fine.

When Joyce told him about her "premonition," he was almost angry. The truth was it frightened him a bit, like the *Ferry Inn* business. He offered no word of comment.

"For Heaven's sake, darling, you must admit it's strange!" she said. "The time coincided exactly. How do you explain that?"

"Chance," he grunted. "Pure chance, that's all. Now please, Joy-girl, let's talk about something else." She never mentioned it again, and nothing like it has happened to them in the years since.

*   *   *

While his hands were healing they sent him on a liaison job with the Merchant Navy. At Southend, on the Thames Estuary, he joined a convoy of sixty ships, bound for Halifax, Canada. He was to sail with it as far as the Firth of Forth, where it was to be joined by another twenty-odd vessels for the hazardous Atlantic run. They let him take young Ronnie Jarvis with him. The four-hundred mile voyage took nearly five days and was not without excitement. . . .

The two pilots were accommodated on board the *Brittany Coast,* a sturdy merchantman of some 1,400 tons which marched in the front rank of the long, slow-moving procession. They spent most of their time on the bridge with the ship's master, Captain Jones, and the commander of the convoy, Commodore Bennett, a dapper, red-bearded man who never seemed to sleep. Bennett explained the elaborate systems of defense, U-boat detection, communication between ships and navigation. Every man in the convoy stood wartime watches—four hours on, and four off. Tuck began to realize how very much tougher, more dangerous and complicated a seafarer's life was nowadays than when he was on the old *Marconi.* . . .

Commodore Bennett had some disparaging things to say about the R.A.F.'s ideas of providing air-cover for convoys.

"Once out of sight of the shore," he growled, "British fighters are about as plentiful as importuning mermaids. And when one *does* show up, unless we're being walloped by a pocket battleship or bombed by a couple of Jerry squadrons he gets bored after a few minutes, waggles his wings and buzzes back home to tea."

"You'll be off Norfolk soon," Tuck said confidently. "I can promise you plenty of cover there—my boys, from Coltishall, look after that stretch."

"I doubt if they'll even hear about us! You see, there are mines in the area, so we won't be hugging the coast. I regret to say it's my experience that when we're out of sight, we're out of mind. I hope you chaps will see that for yourselves and make a report that'll shake Air Ministry."

The Commodore was right. As the convoy crawled northward, with the Norfolk coast just over the western horizon, they saw only two Hurris—and these at different times, in the distance, scuttling about other business and ignoring the column of ships. Late in the afternoon the sea suddenly grew calm as glass, and a mist began to form. Tuck, on the bridge, fancied that he heard a faint thrum of engines and, peering ahead, spotted an aircraft crossing the convoy's path, far ahead, very low over the water. He saw it only for a couple of seconds before it vanished into the gray east. Not a Hurricane this time. . . . He turned to Bennett.

"Sir—there's a Ju. 88 out there! If he saw us—and I'm sure he did—he'll turn right around and come in from the landward side!"

"Thank you very much," said the Commodore, reaching for the warning button. During the next thirty or forty seconds he barked out orders that started the whole convoy into a turn to port and sent every man on each of these sixty vessels sprinting to his action station.

The Junkers, engines wide open, came howling out of the mist to port at about forty feet, its forward guns spitting their streams of poison. The German pilot had selected as his victims the three ships in the front rank —rightly surmising that the convoy's Commodore would be on one of them. Tuck glanced down from the bridge and saw a machine-gun of Indian Mutiny vintage in the center of the well-deck—unmanned. He flung himself down the metal steps and ran to it.

The first bomb raised a spout of water scant feet from the stern of the little tanker to port, then the raider was thrusting straight at the leader. By the time Tuck opened fire, the bomber was already receiving attention from a variety of light A.A. weapons on several of the ships.

The old gun vibrated and kicked so much that its mounting post threshed to and fro, the deck rose and fell, and Tuck's bones jarred painfully. But he kept on firing, and felt reasonably sure that he was scoring hits around the nose. Other gunners were hitting too,

but the Hun came on remorselessly through the storm
of metal, flying straight and level and quite steady,
as though supported by the cone of tracer. And now
the German's shells were punching holes in the hull-
plates . . . knocking jagged splinters out of the life-
boats on the for'ard davits . . . blowing in whole
sections of the superstructure.

Tuck, took his eyes from the sight as the Junkers'
second bomb detached itself and came forward and
downwards with amazing slowness, like a great black
beetle too bloated to fly. Its forward trajectory was far
from spent when it hit the water about twenty yards
off the rail, so it bounced back up, fifty or sixty feet,
and came sailing on again. Perfect skip bombing!

He returned his attention to the sight and gave the
Junkers a last burst as it thundered over the mast-
wires, then he wheeled round and fell on one knee,
arms instinctively raised to protect his face. The bomb
described a leisurely arc clear across the well-deck,
maybe fifteen feet above the planks, narrowly missing
the sheer front of the bridge. It smacked the sea's face
once more about twenty-five yards off to starboard, and
bounced up again. A lower, shorter arc this time: on
the downward slant it smashed into the hull of the
third vessel of the leading rank, an 800-ton freighter.
A red, roaring volcano dead amidships. The little vessel slowed, listed, and immediately began to settle amid
clouds of steam and smoke.

Jarvis was still on the bridge, staring at the founder-
ing freighter in sickly fascination. The Commodore
turned to him and bawled: "Where the Devil are
those bloody pals of yours now?" Then he rattled out
orders for some of the other ships coming up behind to
pick up survivors from the freighter. (They heard
later that none of the crew were lost, though several
were injured in the explosion.)

Tuck found that in firing the gun he'd burst all the
blisters on his hands. In the excitement he'd completely forgotten those burns and felt not a twinge of pain.
The ship's doctor, applying fresh dressings, explained
that the gun hadn't been manned because it had been

found faulty and reported "highly dangerous"—in fact, it was only through some oversight that its ammunition clip had been left on!

The Commodore's anger was shortlived, but he continued, in a fairly good humored way, to gripe at the pilots about the absence of fighter protection. He maintained that the Merchant Navy, in reality a "civilian organization," was many times more efficient than any of the three armed services. In the circumstances they found it hard to argue in defense of the air force. But the next night provided an incident that turned the tables and made the Commodore's face red as his beard.

They were smooshing through flat, glossy water somewhere off Grimsby about midnight when a powerful motorlaunch roared alongside. Tuck and Jarvis, hearing the whine and bumble of its two huge engines being thrown into reverse to slow it down, raced from their cabin to the bridge in time to hear an affected voice hailing them from the darkness.

"Ahoy, there! Anybody awake up there?"

Bennett lifted his megaphone, opened a sliding pane in the side of the bridge and answered: "Ahoy—what's the trouble?" Tuck peered out over the Commodore's shoulder: all he could see was a dim, low shape standing off about fifty yards from them. He couldn't resist saying: "He hasn't identified himself. Why hasn't he flashed the letters of the day?" But Bennett ignored this.

That lazy, affected voice came drifting over the water again: "What ship are you, sir?"

"S.S. *Brittany Coast,* Convoy Easy-Charlie 38." No hesitation in this answer!

"Rightie-ho, jolly good!" the voice called back blithely. "Thank you, sir. Good night!" The launch's engines whined and bumbled into full life again, and the low, rakish shape started to draw away into the blackness. The Commodore didn't even have time to close the sliding pane before the fusillade of shells and bullets struck the merchantman. Everybody on the bridge dropped flat as the tracer licked close overhead. For the best part of two minutes the ship was raked from stem to stern. Deck gear was smashed, the superstructure took a further beating and a lot more holes were punched in her plates—fortunately all above the waterline. It seemed a sheer fluke that the only casualties were minor injuries from flying glass and splinters of wood and metal.

The attacker, of course, had been an audacious German E-boat. With a captain who spoke in a perfect Oxford accent. After that, Bennett stopped criticizing the efficiency of the R.A.F. Tuck and Jarvis made a point of calling to each other frequently in his hearing: "Rightie-ho, jolly good!", and for the rest of the trip went around humming or whistling the old-time ballad: "I Hear You Calling Me." But when they left the ship in the Forth they were perfectly serious in promising the Commodore that they'd do all they could to get Air Ministry and H.Q. Fighter Command more interested in the idea of providing standing fighter patrols, at least during daylight hours, over convoys which, though out of sight of the coast, were within reasonable range.

*　　*　　*

At the end of June Hitler attacked the Soviet Union. Now the British fighter pilots understood why so few Messerschmitts had been rising to meet them over France and the Low Countries in recent weeks—obviously a large number of Luftwaffe squadrons must have been transferred to the east.

Tuck was astounded: he would never have believed that even the megalomaniacs of the Nazi regime could commit such a fundamental blunder. They'd failed to conquer the small isles of Britain, yet here they were turning their backs on the west and deliberately picking a fight with a whole continent—potentially one of the greatest powers on Earth!

He felt a surge of excitement: thanks to Hitler's folly, the British were no longer alone, they had acquired a great and resourceful ally! Russia might take a sorry beating in the first few months, but he was certain that she could never be conquered. He remembered the story of Peter the Great . . . Napoleon Bonaparte's disastrous retreat . . . the failure of the Kaiser's armies even when the Czarist forces had broken. . . . He imagined massive formations of Cossack cavalry sweeping over the steppes . . . millions of peasants and workers and tribesmen from remote eastern areas, vast rivers of humanity, all being armed and drilled and welded into a colossal army. . . . And he thought with glee about the terrible Russian winter, which Aksinia had described to him so often as "the time when everything turns to white iron, when all the wild creatures vanish from the Earth, and even in the great towns and cities people feel remote from the rest of the world."

If only the Ruskies could hold out until the winter. . . . He was convinced they would. In the meantime, every blow that was struck in the west reduced the amount of pressure Germany could apply in the east.

His pilots shared his optimism. It was good not to be alone in the fight any more.

But soon Prosser was whipped away to command his own squadron. (It was becoming impossible to keep a good flight commander—only by blatant wire-pulling did he prevent Cowboy from going. Despite the offers of promotion, the Canadian had no wish to leave 257; he once told the Station Commander, "Bicycle": "I wouldn't feel nearly so good, or fly so good, if I didn't have Bob Tuck out there in front—so you see, sir, I'm of more use right here.")

A letter arrived from Buckingham Palace giving Tuck details of the investiture at which he would receive the second bar to his D.F.C., and enclosing two tickets for the spectators' gallery. He destroyed the tickets and went alone.

The King greeted him warmly and said: "My daughters are always asking about you, you know. I will be able to tell them I have seen, and talked with you to-day. They will be very thrilled. They cut your picture out of the papers, and I may tell you that whenever there's anything on the radio about an air battle they come and ask me 'Was our Mr. Tuck there?' So you'd better watch what you get up to!"

In mid-July Sholto Douglas gave him command of the Duxford Wing. It was a wonderful appointment, but it was hard parting from the "Burma" boys. It was exactly ten months since that evening when he'd walked into their mess, looked at their sullen faces and doubted if he'd ever manage to turn such geese into swans. Then they'd given him September 15th, the day he'd cherish in his memory above all others. And never once since then had they let him down. Needless to say, he expressed none of these thoughts to them, but took his leave in the customary merry manner by throwing a big beer party.

The one great consolation was that he was handing over command to the closest friend he'd ever had—Cowboy Blatchford. It wasn't exactly formal: Tuck drained his last pint, glanced at his watch, smacked the Canadian's shoulder and said: "I'm off—they're all yours, Fat Arse."

"Roger," Cowboy drawled. They clasped hands and their knuckles showed white. "Now that you won't have me to look after you, watch out crossing the road, eh? S'long, Bobbie—keep it in your pants, chum. . . ."

But the last man to shake hands with him before he left the station was the admirable Hillman, his faithful fitter.

"Come back and see us nah and then, won't you sir?"

"That I will."

"I 'ope you won't mind me sayin' this, sir, but you've been a good friend to me." As ever, Tuck shied away from the sentiment. He gave a short, hard laugh.

"Now don't talk nonsense, Hillman—you do your job well, that's all." Then, to soften that a little, very quickly—"Thanks for looking after me."

"A pleasure, sir. Well—all the best, sir."

"All the best, Hillman." He hurried off.

He never saw Hillman again. But the little Cockney was right, they *had* been good friends.

From the very start, all the ground crews had got along splendidly with Tuck. John Ryder, the "instrument basher" who used to strap him in the cockpit, states: "257 was a rather indifferent squadron before Tuck arrived, but in a very short time he had it on top line. Tuck always took a real interest in his ground crews and never had that 'toffee nose' attitude that some of the pilots displayed towards us. He gave us all consideration and confidence.

"To see him shoot clay pigeons was an experience. What an eye he had!—no wonder he had such a large score in the air. Whenever he went to the shooting range there was always a mob of spectators. So far as we airmen were concerned, he was an 'ace' on the ground as well as aloft."

Now, a few days after his twenty-fifth birthday, he became a wing commander, with three squadrons to run. No. 601, formerly commanded by Max Aitken and now by "Jumbo" Gracie, was operating Spitfires Mark V, which carried two 20 mm. cannon and four .303 machine guns. But they were just being re-equipped with the Airocobra, the revolutionary new American fighter which had its engine in the fuselage behind the pilot, and was fitted with a tricycle undercarriage.

No. 56 had just been taken over by Peter Prosser Hanks from Coltishall; they were flying Hurris, but were in process of converting to the new and powerful Typhoon—they were, in fact, the very first operational "Tiffie" squadron.

Lastly, there was No. 12, better known as the Third Eagle Squadron. Every pilot was an American national and a volunteer, and they were quite satisfied with their Spit Vs. Tuck expected to find the Yanks walking around in baseball caps and fancy leather jackets with "Spike" or "Hank" or "Butch" stenciled on the back. But he found they were almost indistinguishable from the pilots of the other Duxford squadrons—they dressed just the same, and they used the R.A.F. slang more than their own native idiom. They didn't even chew gum much.

He missed seeing Joyce in the evenings—and he was

surprised at how very much he missed the quiet comfort of her home. But she wangled weekends off and came to Duxford. Once or twice Cowboy came too, and then it was just like the old nights at Coltishall.

The wing was hell to manage if he tried to fly the Tiffies and 'Cobras along with the Spits. The different types varied in their cruising speeds, rates of climb, turning distances and many other characteristics. He put in several hours on both the new fighters, and liked the Typhoon especially, but for the meantime he decided that when they operated as a "circus" he'd stick to a mixture of Spits and Hurris.

There was no hesitancy about what he'd fly himself —the Spitfire was his true love. And the Mark V was decidedly faster and more formidable than the II.

Nowadays a "circus," or wing formation, was more often called a "balbo"—after the famous Italian aviator General Italo Balbo, who before the war had led vast fleets of aircraft about the world. Lately the Germans had been putting up surprisingly strong resistance against our fighter sweeps, and occasionally dramatic clashes had developed over the other side between British and German balbos.

The first time Tuck took his herd to France, sure enough they found a big chandelier of Messerschmitts hanging up waiting for them. Lots of yellow noses in there. The two formations sparred round, each looking for an opening, then the Germans, who had a slight advantage in height, started peeling off in twos and threes and coming in on the British from all directions, trying to provoke them into splitting up.

But Tuck wasn't ready to break up his formation yet—he'd wait till a lot more of those 109s had come down. He put his chaps into a wide defensive circle. He'd chosen as his number-two one of the younger pilots who needed experience and encouragement. In the turn the kid got a little wide.

"Look out behind!" Several of them were yelling at once. He tightened his turn and looking back over his shoulder saw two 109s flying right up through the long, curving stream. They were not firing. They skimmed adeptly under or over each vic of Spits or Hurris.

"Watch it up front!" That was Prosser's voice.

Tuck tumbled to it now. They were after the leader —they were coming for him. . . .

A moment before the tracer stabbed at him he jinked to the left, like a boxer side-stepping. The fire went wide. As the two would-be assassins streaked past and climbed hard to rejoin their main formation, he got in a quick squirt at the number-two man. The tail disintegrated and the 109 went down, whirring round like a top. The other one reached the sanctuary of the "chandelier."

But as he eased back into position at the head of the wing he saw that his own number-two was missing. (Prosser said later: "He got too wide in the turn— 'way out on his own. The leading Hun got him, once he'd realized he couldn't get you.")

The two balbos skirmished for a few minutes more, and Tuck damaged another one, but a fullscale dogfight didn't develop. Fuel was getting low, so he took the boys home.

Crushing news: Doug Bader was missing. He was believed to have bailed out over France. The mess was hushed that day. The central pivot had just fallen out of an era.

To every pilot in the air force, Bader was a symbol, an inspiration. He was the living embodiment of the stubborn spirit which had beaten back the Luftwaffe's summer assault, and of the Churchillian toughness and resolution which were now the driving force of the mounting air offensive. There was hardly a man on fighters who hadn't said to himself in moments of stress: "Dammit, if a bloke with no legs can do it, *I* bloody well can!" And make no mistake—Tuck was no exception here!

Then the radio announced that the Germans had claimed him as a prisoner. Everyone laughed in relief, and the bar did record business. "My God," Tuck said. "I'll bet he gives them a helluva time!"

They were having a little teething trouble with the Tiffies. Carbon monoxide fumes were coming back from the stub exhausts into the cockpit, sometimes in sufficient quantity to nauseate the pilots. The engineers came and tinkered with the exhausts and said they thought they'd reduced the amount of fumes. To make sure, they fitted in the cockpit of one machine a little box which could measure the exact percentage of carbon monoxide in the air inside the cabin. They asked for a number of test flights: Prosser Hanks did most, Tuck a few. There seemed to be a definite improvement.

One day, towards the end of the test programs, he was walking out to the test Tiffie with his parachute over his shoulder when an airman stuck his head out of the dispersal office and called: "Telephone, sir!" At that moment the ground crew finished their starting drill and the Typhoon's huge Napier Sabre engine exploded into life: if she didn't take off quickly she would overheat. He signaled that he couldn't take the call and continued out to the aircraft, but the airman came running after him and bawled through cupped hands: "It's the station commander, sir. Says it's very important!"

He groaned, dumped his 'chute and started back. At the door of the dispersal hut stood one of his best pilots, a young Argentinian named Dack. This boy had flown one or two of the tests. "Dacky, you take her. You know the drill." The kid nodded, grabbed his gear and hurried out as Tuck lifed the phone.

Group Captain MacDonald wanted to discuss arrangements for night-flying during the coming week. It *was* important, but it could have waited an hour or so. They talked for perhaps ten minutes, then Tuck went out and sat in one of the cane chairs, smoking and looking out over the field.

Out of the hazy blue he saw a Tiffie diving. It didn't pull out. It disappeared behind some trees about a mile on the other side and raised a tremendous cone of flame and smoke.

He rode out with the crash trucks. A big, smoking

crater and a field littered with scraps of metal. Twelve feet down in the brown earth: the remains of the big engine. Of the pilot: only little red lumps, half a shoe, scraps of clothing, part of a watch strap.

A check with control proved it could only be Dack. There had been no enemy activity all day, so he hadn't been shot down. The cause of his death might always be a mystery, because there wasn't enough left of his machine to give the technical experts a clue.

But the Aviation Medicine people solved it. They found a piece of Dack's liver and analyzed the contents: enough carbon monoxide to kill an elephant.

Probably something had gone wrong with the technicians little box. Instead of trapping the fumes it must have pumped out enough to make the pilot pass out.

There but for the grace of a phone call. . . . He got through to Group Captain MacDonald and in a quiet, earnest tone said: "Sir, I want you to know you can ring me up just any time."

15

"Tucky, you're coming off operations for a while. High time you had a breather. Air Ministry wants you to pass on some of your 'gen'* to chaps who haven't seen action. . . ."

The voice on the other end of the line belonged to Air-Vice-Marshal Richard Saul, the A.O.C. of No. 12 Group. Tuck had known for a long time this was bound to come, and he immediately launched into the passionate appeal he'd prepared against this moment. But the A.O.C. was very firm.

"Sorry, it's all been arranged. There's no argument."

All right then, dammit—second line of defense: "But sir, can't you find me something—*anything*— apart from Training Command? I'd never make a bloody instructor—that would drive me off my rocker, honestly sir! Surely I could do some test flying . . . or even——"

"Ease up, Tommy!" A deep chuckling came over the line. "I haven't said a word about Training Command. If you'll only let me get a word in edgeways I'll explain what's arranged for you, and I'll be damned surprised if you grumble!

"Now listen—Air Ministry have picked you to go on a special tour, along with a few other highly experienced chaps. You'll be talking to thousands of pilots—*trained* pilots, gasping for all the latest combat information—and what's more you'll put in a lot of flying on brand new types. Does that sound so bad?"

*Knowledge.

"I . . . don't quite understand. . . ."

"Well, then, let me put it this way: how'd you like to go to America?"

October 1941: the Germans were rolling steadily east towards Moscow, Greece and Crete had fallen, and in Britain food rations were reduced because the U-boats were sending an average of 180,000 tons of our merchant shipping to the bottom every four weeks. Night-bombers still struck at London and the Merseyside, and there were now more than a million women at work in the munitions factories. There was a big drive for "scrap:" park and garden railings had disappeared to be transmuted into tanks and guns, and aluminum saucepans were sacrificed to provide fighters and bombers. Fines for breaking the black-out regulations were getting stiffer, there was a shortage of Scotch whisky, cigarettes and chocolate, and pretty girls were having to do their best with lisle stockings and standardized utility dresses.

After the grim, make-do-and-mend atmosphere of the besieged islands, neutral America seemed at first like one huge, brilliantly-lit and very noisy fairground where everyone was bent on having a "good time and the hell with tomorrow." A land flowing with milk chocolates and honeys in sheerest nylons.

Jittery electric signs, madly gaudy taxicabs, T-bone steaks, all-night cinemas, baseball matches, brand new motorcars, the mingled smells of popcorn and salted peanuts—it was everything Tuck had heard, but never quite believed. He was dazed by the roaring energy all around him.

There were five others in the party and to his delight one of them was his old chum Sailor Malan. Sailor now had twenty-eight victories to his credit— at this time, two or three more than Tuck. The senior fighter pilot was Group Captain Harry Broadhurst, who'd been a squadron-leader before the war. At 36 he was still as fit and fast as any of the youngsters. The standing joke about him was that he'd been flying "since balloonists wore red tights." Like Victor Beamish, "Broady" was never content to sit at a quiet desk with his squadrons singing overhead, and frequently he'd led the Hornchurch Wing into battle. The remaining three were bomber "aces:" Wing Commander Hughie Edwards, V.C., a youthful Australian who'd once bombed a Bremen target from fifty feet; Wing Commander J. N. H. "Charlie" Whitworth, ice-brained veteran and a wizard at long-distance navigation, and Group Captain John Boothman, former Schneider Trophy pilot who had been planning—and sometimes leading—many of the biggest raids on the Ruhr.

Washington. Photographers waiting on the tarmac . . . the British Air Attaché and a horde of United States Army Air Corps officers ushering them swiftly to a fleet of gleaming limousines, telling the reporters: "Sorry, boys, no statements" . . . the ride through the city with motorcycle escorts . . . the whole convoy going out of its way so that they could take a quick look at the Capitol and the Lincoln Memorial . . . the centrally-heated luxury of their individual suites at the *Lee-Sheraton* . . . a small but genial reception in the evening, attended by a few congressmen and senators with their ladies, officials from the Bureau of Aeronautics, army and navy pilots, people from the British Embassy. . . .

And next morning, to the Pentagon. The surprising military formality of the Americans—grave faces and jerky, upswept salutes every time an introduction was made . . . then, in startling contrast, their relaxed, easygoing familiarity once business had been done. Long, wearying sessions of interrogation with the technical types and staff officers . . . finally an interview with General "Hap" Arnold.

"Glad to welcome you here, gentlemen. You can help us, and I hope we can help you." A fierce, carefully controlled, solid man with silver hair and eyes that seemed to drill into you. Only a few questions, but every one of them shrewd, then: "You're going out on maneuvers now. You'll see everything we've got, *and* the things we haven't got. The time for us to talk will be when you get back."

America had reached maximum generosity in her interpretation of neutrality. Earlier in the year the government decided it was pointless to send millions of dollars worth of lend-lease material across the Atlantic unless they could be sure it reached its destination, and so their naval and air forces had been ordered to search the whole of the western Atlantic for Axis vessels. American escorts were provided to take lend-lease convoys two-thirds of the way to Britain, and United States troops had joined British forces in Iceland.

Then, only a few weeks before, President Roosevelt had made that fateful declaration: "American naval forces and American planes will no longer wait until Axis submarines, lurking under the water, or Axis raiders, working on the surface, strike their deadly blow first. . . . Let this warning be clear: from now on, if German or Italian vessels of war enter waters the protection of which is necessary for American defense, they do so at their own peril. . . ."

Tuck had the distinct impression that the vast majority of the American officers he spoke with were convinced that their country's entry into the war was inevitable—though not one of them ever expressed that opinion in so many words. With them Japan was as big a menace as Germany and Italy. Most of the senior officers seemed consumed by a sense of urgency, bemoaning the fact that a drawback of the democratic state was its habit of always getting caught unprepared.

They all knew it would take some kind of crisis, an act of greatest provocation, to carry public opinion over the brink and drown out the Isolationists: they had no idea what shape this might assume, they just felt sure something was bound to happen—and it was clear that most of them thought "the sooner the better now."

South Carolina. The R.A.F. delegation was split up, and the three fighter pilots joined Air Corps units for the combined maneuvers. Sailor went to fly P.38 Lightnings with the "Blue Forces" in the mock war, while Tuck joined the "Reds"—who turned out to be the First Pursuit Group, several squadrons equipped with P.43s and based at Spartanburg. Harry Broadhurst accompanied Tuck, but his job was to be mostly on the ground—consulting with the American commanders and technical representatives, and in particular keeping an eye on ground control procedure.

As their guide, companion, "guard and interpreter" for their stay in the United States, they had a short, tough Westerner named Major Wynn. He was a pilot, but by peacetime standards too old for fighters. His legs were slightly bowed and he had a slow, deep drawl. By an almighty coincidence, (though obviously not without good reason), his nickname was—"Cowboy." He soon proved himself a man with easy authority, and brilliant flashes of fun. He looked after them with avuncular concern—as if they might get lost, or fall into dreadful pitfalls, if he let them out of his sight for long.

The P.43 was a small, stubby fighter armed with

machine guns, two of which fired through the propeller's arc on a synchronization system which reminded Tuck of his Gladiator days. It was slower than the Spitfire and Hurricane, but very solidly built.

The pilots were eager, industrious and intelligent. "The thing that impressed me most about them," he says, "was their great capacity for self-criticism. They used to bawl each other out for mistakes made in the air. Several would pile into one poor sod who'd boobed, but I didn't once hear a recipient make an angry retort.

"Their flying was of a very high standard—precise, neat and quite flashy at times. But right at the start I found one serious omission in their work: very few of them had ever been up above fifteen thousand feet, though the P.43 could operate at twenty-five thousand and even higher. They just didn't appreciate how important an advantage height was in modern air-fighting. It shook them when I told them about dogfights at over thirty thousand!

"I soon found why they were so reluctant to climb—they weren't properly equipped for high altitude work. They had cumbersome and ineffectual oxygen apparatus—a sort of rubber lung which hung down over the chest, pulsating all the time and getting in your way every time you leaned forward or turned your head. When you rose into really cold air, the damned thing used to fill up with ice crystals—because your warm breath would condense on the concentric tube inside the bladder gadget and freeze solid. After a few minutes you had to keep squeezing it with your hand or it would just stop inflating and you ran the risk of passing out through lack of oxygen.

"I told Cowboy Wynn the thing was a washout, and I dare say he conveyed that opinion to the proper quarters.

"I dunned into them the rule 'up-sun and bags of altitude,' but it was a while before they got over their height-shyness."

Other, only-to-be-expected faults: stylish but much too cramped formations, no r/t discipline, not nearly enough night-flying experience. But their gunnery was quite brilliant—especially on ground-strafing—and they had plenty of guts and enthusiasm.

Bright, boyish faces, with steady eyes and sun-scarred noses, cheeks and lips—no different from the faces of the youngsters back at Duxford or Coltishall or Hornchurch. Squatting on the grass listening to him, in those familiar postures which seem to be peculiar to flying men at such moments . . . wearing the same earnest frowns, thinking and working with the same quiet ardor. Pilots were a world-wide fraternity: he supposed they looked, thought and behaved not much differently in Germany or Italy.

Certainly Tuck was greatly taken with the men of the First Pursuit Group—Schaefer and O'Brien, Higgins and Perez, MacDonald and Greenbaum. . . . He would have liked nothing better than to lead them out across the Channel one fine day and into a bunch of Messerschmitts. Yes, even with P.43s. . . .

Wynn kept them amused with stories about his early barnstorming days. Once he'd belonged to a famous airobatics team, and he claimed he used to pick up handkerchiefs with his wingtips while flying upside down—"and they were *ladies'* kerchiefs, little things not more than six inches wide." It could have been true, but Tuck pretended to disbelieve him—because then Cowboy got very excited and swore in a comical, colorful style.

The British officers and their hosts were invited to a staff party at Spartanburg's fine, modern hospital. It was supposed to be a "dry" affair, but the Coca-Cola bottles were refilled behind a bed-screen with rye whisky, gin and brandy while the Matron looked the other way. It got a bit hectic after an hour or two, with two of the prettiest nurses performing a can-can they'd done at a recent Orphans' Benefit—apparently forgetting in their excitement that on this occasion they lacked the standard protective clothing.

When at last the party broke up everybody was fairly aglow. In the hallway a pack of the nurses, armed with surgical scissors, suddenly rushed at Tuck, pushed him against the wall and operated on his tunic, deftly removing his wings, medal ribbons and every button.

His protests weren't very serious, for they paid for their souvenirs in a manner that kept his lips occupied. Luckily he'd brought another uniform on this trip.

The fighters of "Blue Army" and "Red Army" tangled frequently above the rolling, scrubby Carolina country. Sailor's Lightnings were faster, and had a better rate of climb, but Tuck's P.43s scored with one or two nicely timed ambushes out of cloud, and made great capital out of their ability to turn in a much smaller radius. Both packs hunted at well above twenty thousand—though, as it turned out later, Sailor had had even more trouble than Tuck in persuading the "Blue" boys that it was really necessary to climb into the thin, chilly air!

The "Red" formations did very well in strafing attacks on the "Blue" land forces. But at the end of the maneuvers nobody seemed to be sure which Army had won. It didn't matter—a lot of improvements had been made on both sides.

Before they left the area, Wynn arranged for him to fly to another airfield, Charlotte, and talk to the pilots there about the results of the exercise. He went alone, in a P.43. His talk was warmly received, and he was rewarded with a long, leisurely lunch and a good deal of liquor and levity. Then one of the senior Charlotte pilots offered to fly with him most of the way back to Spartanburg—to save him the trouble of navigating from maps folded across his knees, always an irritating business in a cramped cockpit.

The day was dust-dry and shimmer-hot. The two little fighters crawled over the egg-brown landscape, dwarfed by the vastness and the sameness of it. After the patchwork quilt of England, it seemed to Tuck featureless and empty as the Sahara. He'd had trouble picking up landmarks on this morning's flight to Charlotte, and now he was grateful for this escort who'd traveled between the two airfields so often he knew the route by heart.

He sat back and rested his head against the rubber pad, thinking how good it was to be able to fly at his ease like this, without having to keep constant vigil for the enemy. Then he spotted two tiny, glinting flecks high above, hurtling straight down through the sunlight. Tuck and his companion were about to be bounced from a very great height—by a couple of scheming youngsters in P.43s, anxious to show this "Limey" they'd taken his teachings to heart. . . .

He barely had time to roll on to his side and pull hard away before the "attackers" got within range. There followed a vigorous, fifteen-minute dogfight in which they turned and twisted, dived and climbed, rolled and bunted, everywhere from twenty-thousand down on to the deck and back up again. The two assailants ignored their compatriot and concentrated on Tuck, trying all they knew, but not once did they come near to getting on his tail.

Then, with that familiar but still amazing suddenness, the scrap was over and Tuck found himself all alone in the sky, breathless and chuckling from the rush and fun of it. Then the chuckling died as he realized he'd lost his guide-plane, and had no idea where he was. In every direction the brown land stretched away to the horizon, unbroken by any large town, railway junction, or distinctive topographical feature.

Well, not to worry. There was a set of procedure for this, and in this perfect visibility he'd soon pick up a landmark and pinpoint his position. He climbed to ten thousand, unfolding his maps on his knees. On leveling out, he made a wide circle, inspecting the terrain.

All glitter and glare. The fierce white sunlight seemed to be drumming on the canopy. Despite his tinted glasses, it made a splintering dance of everything below.

Somehow the aircraft itself didn't seem to be moving: it was stationary, floating in space while the landscape beneath unrolled slowly, silently, like a worn carpet. Nothing stood out. No single hill or river asserted itself, no valley or fault commanded special attention. Every feature had its imitators. Impossible to identify anything on the maps.

Very well, then: fly in one direction for a few minutes, bound to come to a road or a railway or a town. . . .

He throttled back to conserve gas and flew south for ten minutes. He came to no such things. The country didn't change. Now, for the first time he began to comprehend the vastness of America! There seeped into his mind a dreadful thought: what a bloody fiasco it would be if, after serving six years in the air force and logging over 1,000 flying hours without once "writing off" an aircraft through his own error, he was now to run out of juice and wreck this machine—property of another government—in trying to force-land it in this rugged countryside!

Tuck—"the Flying Fox," the Spartanburg boys called him—getting himself lost in perfect weather, smashing up a kite thousands of miles from the nearest German. . . . It was a pitiful prospect—savagely humorous, grotesque, tragic. It mustn't happen: no matter what, *this plane must not be scratched or dented.*

Now he flew east for ten minutes. No change, no dominating peak or pass or plain. There should be airfields out here: he scoured the distance for the gleam of smooth runways—the black cross that would be his salvation, the oily star that would guide him to safety.

Nothing.

He tried the radio, calling on every channel, one after another. No answer. His own nerves seemed to be crackling like static.

He turned north. Fuel getting low now.

Christ, what a mess!

And then he picked up the highway—a narrow ribbon winding across this big, blank hinterland. A lifeline. He clung to it, following its crazy curves through foothills and ravines, letting it pull him back on to a roughly easterly course. Must lead somewhere—sooner or later it was bound to cross a river, or reach a small town, or join another road. . . . It must meet up with some other feature, and the combination of the two—the pattern they made—would give him a "fix" on the map.

Many minutes passed, and many gallons of gas were

consumed, before he saw the little town. It was just a short row of buildings on either side of the highway, but it was as cheering a sight as ever he'd seen. He circled it, fairly low, and fiddled with his maps, unfolding them section by section, striving to identify the place. He identified it five, ten, fifteen times over—it could have been any one of a score of hamlets flanking a dozen different roads in the state. . . .

The disappointment made his throat lock, tremulously tight. He would have to face up to it: fuel was very low, and he must now find a place to put her down.

Beyond the town the road seemed to run more or less straight for about 2,000 yards. That would be his runway—he could depend on it being hard and fairly smooth. In the town he'd get straight on the telephone to Wynn. There might even be a chance of getting some high-octane fuel flown over so that he could fill up and take off again. . . .

He throttled back and made his approach. The road was just wide enough, allowing for slight drift or swerve. Just as he checked on the stick and began to bring the nose up for the touchdown, round the bend far ahead a huge, red sedan came speeding. He hadn't seen a single vehicle all the time he'd been following the highway—what a helluva moment for one to show up!

He added a little power, letting the fighter float along a few inches off the ground at a mite over stalling speed, watching the car coming at him. The bastard wasn't going to stop—probably he thought the plane was just stunting, and was determined not to let some damned young flying fool scare him into applying the brakes. . . .

Tuck cursed, added throttle and out of spite held the plane down till the last second, pulling up hard to clear the car's roof with his wheels. By that time the driver's nerve had cracked, and he was skidding broadside in a storm of dust.

Smartly now—fuel gauge reads only a shade above zero!

But twice more he brought her in only to see a car come round the bend ahead and race towards him.

God, where was all this traffic coming from so suddenly? On the third occasion, convinced that he didn't have enough fuel left to get him round another circuit, for one insane moment he was tempted to fire a burst at the oncoming vehicle and blow it off the road.

Rigid, drenched in sweat, expecting at any instant to hear the rapsing cough of the engine's dying, he coaxed her up again. Four hundred . . . five . . . six . . .

He leveled off gratefully at eight hundred, and made a tight circuit. Then he decided he couldn't risk the road again—the fuel needle was actually *below* the zero mark, resting on the bottom of the gauge!

It would be a miserable epitaph for a Battle of Britain pilot—"Killed in a collision with a Buick coupe!" And that thought reminded him of something he'd noticed on his previous approaches: near the end of the town, just off the road beside a small church, there was a cemetery. It appeared deserted. Its ground was very flat, and down the middle ran a broad driveway, maybe 600 yards long.

Could he get her in? Could he bring her to a standstill in that distance? He wasn't very hopeful—but at least he could be sure he wouldn't meet a speeding motorist in there!

At the far end of the driveway there was a tall, very solid-looking brick wall. If he ran on and crashed into that he'd almost certainly break his neck. Ah well, there looked to be a few vacant plots. . . .

His wheels touched in the first five yards. But the driveway was loose gravel, deep and soft as a shale beach. The moment he started to brake, the wheels sank into it, the nose dipped and the tail kicked up. He had to snatch the stick back, hard, to prevent the propeller tips digging in. It was fortunate that the little P.43 still had a bit of elevator control even at this speed.

The gentlest touchings on the brakes—first one, then the other. Several times she gave a frightful lurch and he felt sure she must go over on her nose, but somehow he just managed to hold her. Now the gravel was pattering furiously on the undersides of the wings and fuselage He held the stick right back in the pit of his stomach, and eased on both brakes at once. The tail bucked once more, then dropped back with a deep, truculent crunch. The fighter rocked and rattled to a stop about twelve yards from the brick wall.

As he swung his legs out on to the wing, a tall, cadaverous man in denim overalls and a wide-brimmed straw hat emerged from among the headstones and shrubbery. He came right up to the edge of the gravel, leant on his hoe and stared at Tuck and the P.43 for a long time. His face was long and infinitely mournful. Suddenly he turned his head, spat out of the corner of his mouth and called in a high, lazy voice: "Hey Luke —young fella heah with an airplane!" Another, older gravedigger, similarly dressed, ambled up, gazed stonily for the best part of a minute then said: "Well dang me, so theah is. . . !"

The fighter had to be dismantled by retrieving crews and brought back to Spartanburg by road. Nevertheless, Tuck felt his honor had been preserved—the machine hadn't suffered a dent or scratch. They kidded him about it, of course, but there was no official inquiry.

*   *   *

Of all the technical brains they met in America, Tuck was most impressed by Alexander P. de Seversky of Republic Aircraft—design genius, author of *Strength Through Air Power,* man of vision and courage.

Seversky—small, sharp-featured, silver-haired—had designed the P.43, but now regarded it as completely obsolete. He drove the two British pilots out on to the tarmac at Farmingdale, Long Island, and with quiet pride showed them the prototype of his new brainchild, the P.47 Thunderbolt. It bore a definite family resemblance to the P.43, but it was much larger, betterarmed—and, he said, considerably faster.

Without hesitation he consented to let each of them fly it—though he did say, smilingly, "Take it easy, won't you?—it's the only one we've got!"

Tuck took the prototype up whenever they'd let him. "It had a very good rate of climb, its ceiling was high and it made a rock-steady gun-platform—and those were the qualities we'd hoped most of all to find in a new American fighter," he recollects. "It lacked nothing in speed, either, but it was very heavy to handle and not nearly so maneuverable as any of our own fighters. Quite definitely it didn't have the performance of the latest marks of the Spitfire.

"And yet, I was satisfied that this big, strong, hardhitting kite could take on the Messerschmitts with confidence. I told Seversky just that, and I think he was very pleased.

"I must say he really was a wonderful little chap. He knew more about the air tactics being used in Europe than anyone else that we met during our whole visit. He had only one idea in life: to provide his country with first-class fighter planes, so that she need fear no other power on Earth. To that end he worked unsparingly, and successfully."

Seversky introduced him to a number of New York businessmen, and in the evenings he was invited to many parties. He met a lot of celebrities at those affairs, but only three of them struck him as "honest-to-God people"—Wall Street financier Calvin Bullock, actor Edward G. Robinson, and Miss Rita Hayworth.

Bullock he considered a wise and gentle old man, deeply religious, grieved by the human suffering of war, yet realistic in his attitude. He still hoped America could stay out of it, contributing to the struggle with her vast industrial power rather than with the blood of her young men. But he faced the fact that her entry into open hostilities was growing increasingly probable every day.

His greatest joy and pride was his "museum"—a collection of relics of Napoleon Bonaparte and Admiral Lord Nelson which he kept on the top floor of his premises at Number One, Wall Street. He was pleasantly surprised when Tuck, viewing these treasures, displayed some knowledge of art objects and antique silver—Nunkie's doing, of course. It transpired that many years before Bullock had been a regular customer at Ackermann's Gallery in Bond Street, and he remembered Nunkie very well indeed.

Before Tuck left New York the millionaire gave him

a beautiful, gold-mounted replica of Lord Nelson's sword. (On returning to Britain he hung it on the wall of his room, and always cleaned it himself once a week. It was stolen from there within a few hours of his being reported missing over France.)

Edward G. astounded him: the tough, drawling gangster of filmdom was in real life a brilliant conversationalist, a scholar and a linguist! They discussed European languages and soon got on to Russian history and literature. The actor shared Tuck's belief that very soon now, when the snows deepened in the Ukraine and points east, the Wehrmacht was in for a nasty shock. They drank to that in very fine brandy. Several times over.

Miss Hayworth exhausted him: she eyed his long, lean frame and said: "You, my friend, ought to be a dancer!" Laughing away his anxious protests, she wafted him on to the floor and proceeded to instruct him in the art of jive. At the end of an hour, to his own amazement, he could do several very fancy steps. But he was drenched in perspiration and his legs, feet and arms ached as though he'd been saber-fighting, while Miss Hayworth remained fresh and full of pep.

"Not bad for your first lesson," she judged. "The trouble with you Englishmen is you're all a bit self-conscious, you never really let yourselves go. Never mind, next time we'll try ballroom style." Next evening she took him to Radio City to see her latest film, "You'll Never Get Rich!"—in which she danced with Fred Astaire!

On two other evenings she took him on tours of the city in her chauffeur-driven limousine, finishing up somewhere in Greenwich Village. She didn't make a dancer of him—he still hated it—but she was fun to be with. They drank a lot of wine, laughed a great deal and hardly ever mentioned the war. He told her about Joyce and that was the only time she said anything that was wholly serious: "Well, I certainly hope everything turns out right for both of you." It wasn't meant to sound grim; it was said quietly, with the greatest sincerity.

* * *

Wright Field, Dayton, Ohio. Intensive flying—on P.43s, Airocobras, Kittyhawks, Tomahawks, various versions of the Lightning, and Mustangs fitted with different types of engine. He liked very much indeed the Mustang which had a Packard-Rolls power plant—a Merlin engine designed by Rolls-Royce in Britain, now being produced under license by Packards at Detroit. He got a big thrill out of a visit to Kitty Hawk Beach, where the brothers Orville and Wilbur Wright had made the world's first airplane flight.

It happened that Wright Field had the only Spitfire in America—a Mark V. Unfortunately almost every pilot in the Air Corps had had a go on her and, like a car that's had too many drivers, she was sadly the worse for wear. But when, on his last day there, his hosts asked him to "put on a show" with her, he felt it would be letting the R.A.F. down if he refused.

"She was very tired, very sloppy—she'd had the guts caned out of her all right. But even so, in the air she was still sweet to handle. I managed to put on a few twiddly bits for our hosts."

His "few twiddly bits" included a progression of loops, upward rolls, rolls-off-the-top, and inverted flying at low altitude—he came out of one maneuver and right into another, so that the aircraft never appeared to stay straight and level for more than a very few seconds.

"One thing they seemed to appreciate was a favorite trick of mine. I came in over the hangars at around 800 feet fairly slow, at about 135, and then did a stalled flick-roll. It was really very easy, if you knew the Spit well. She literally stalled around her own axis and fairly whipped round.

"Mind you, you had to be careful that you started to correct and check the roll at the right instant, otherwise she'd stay stalled, and go into a spin. When you did this properly, you wouldn't lose more than a couple of miles an hour—and she'd carry on at exactly the same height as if nothing had happened. It was a very pretty thing to watch from the ground."

The Americans were chary of airobatics at low speed. This was largely due to the fact that their fighters were apt to stall and spin without any kind of warning, whereas the Spitfire gave her pilot ample notice by shaking and rocking gently several seconds before she was liable to drop a wing. They said they'd never seen anything like that before, and from the way some of them looked at him it was clear they thought he was a case for the psychiatrist—"Messerschmitt happy" was a phrase he heard somebody murmur. . . ."

Back to Washington, where Sailor joined them. "Hap" Arnold saw them again, and in this interview they stressed the need for more high altitude work, and, above all, more night-flying. All the British pilots were deeply concerned about the Americans' apparent belief that they should confine all their operations to daylight hours—even the bombers were "bedding down" at dusk. (This policy was not changed after America's entry into the war, with the result that her daylight bomber groups suffered appalling casualties. However, since R.A.F. "heavies" operated for the most part by night, the partnership had the virtue of keeping pressure on the enemy right round the clock.)

The General listened intently, fixing each speaker with those steady, piercing eyes. When he asked a question it came out crisp and quick, nevertheless you had the impression that every word had been carefully weighed beforehand. The man emanated a dramatic atmosphere of urgency and determination.

Later there was a cocktail party, during which Tuck praised the shooting of the American pilots in the Carolina maneuvers. The General said: "Americans are natural marksmen. We still do a lot of hunting. Lot of my boys grew up with rifles in their hands, providing some of the family's meat."

Tuck, who had a nice glow on, said: "General, a man's eyes often show whether he's a good shot. I should say you're pretty handy with a gun yourself."

It was the first time they'd seen "Hap" really smile. "I'm reckoned fair. And I hear tell you're not bad, either. What d'you say we shoot it out, Commander? Dawn tomorrow suit you?"

They settled for 8 a.m. The General's car picked up Tuck and Wynn at the *Lee-Sheraton* and whisked them to a big air force range on the outskirts. It seemed that news of the "duel" had swept the capital; hundreds of officers and men, and a good many civilians, were waiting at the butts, and bundles of dollars were being brandished and odds called. Everyone wanted to back the General. Cowboy put a total of fifty bucks on the Englishman, made up in small bets, mostly of five and ten.

"Hap" had two beautiful, ivory-handled Smith and Wesson revolvers in a leather case.

"The thirty-eight, like the forty-five," he said, "is a good old American weapon. Choose, m'boy." Tuck chose, thoroughly dismayed—he'd taken it for granted they'd use rifles. He hadn't done any shooting with a revolver in many months.

They tossed a coin and it was the General's turn to start off. He had the whole thing worked out: "We'll fire twenty-four rounds each, Commander. First of all on normal application, twenty-five yards, then we'll come back to fifty. After that rapid fire at the same range. All right by you?"

"Fine, sir." Might as well *sound* confident. . . .

He watched very carefully as the General fired his first six shots—noting the straightness of his back and shoulders, the way he kept his chin tucked in, how he brought his shooting arm up slowly and smoothly, with the elbow very slightly bent. This was all very correct, drill-governed—altogether different from the kind of revolver shooting he'd done in England (mostly just blazing away at old tin cans!) He was trenchantly reminded of another "duel" long, long ago: he'd felt just like this as he watched "Portuguese Mick" hurling knives at the fo'c'sle wall in the old *Marconi*. . . . Now, as then, he studied his opponent's every move and posture, and copied it when his own turn came. The result was that he won by two points.

"Hap" shook his hand and said: "Nice going, Commander. First time anyone's licked me with this brace of pistols." Tuck moved to put the revolver back in its leather case, but the General stopped him. "Like you to keep that one. Souvenir." He snapped the lid shut, slung an arm round Tuck's shoulders and said:

"And now, once Major Wynn's finished stuffing his pockets with dollars, let's go have some breakfast."

Twelve hours later the six R.A.F. men left Washington. Cowboy was deeply affected as he shook hands for the last time. The graying little Westerner had cared for them these seven weeks like a kindly uncle, and never once had there been a serious hitch in his arrangements.

"Wish I could take you with me," Tuck said. "I could use a damned good adjutant."

"Godammit!" cried Wynn. "I can sit on my fanny and fill up lousy forms right here! If I come over there you'd better have a Spitfire for me. You could use a guy with *real* experience. Ever tell you how I used to pick up 'kerchiefs with my wingtips. . . ?"

*   *   *

On December 1st, 1941, having delivered to Air Ministry an exhaustive report on his U.S. trip (including an evaluation of each American fighter he'd flown), he took over command of the Biggin Hill Wing. Now he had *five* squadrons to manage, all flying Spitfires, Mark IX or V: the "First Canadian," (No. 401 R.C.A.F.), No. 72, No. 91, No. 124, and No. 264 (night-fighters). *Lord of sixty pennants. . . .*

Exactly one week later units of the Imperial Japanese Navy attacked the United States base at Pearl Harbor, and Roosevelt declared war on the three Axis powers.

Tuck fished in his kit for a bottle of Scotch he'd bought in Washington—the stuff was getting rare as gold in Britain—and placed it, seal intact, on the little table in the middle of his room. Good to have a fine liquor on hand, now that he was expecting friends.

16

He looked down at the man in the water, a tiny, pitiful bundle tossing amid the big, killing combers, and compassion was a bitter strain in his chest. The sea took so many, the blind, indifferent sea . . .

It was late afternoon and the December sunlight was thinning away. Frost rimed the wings, and the cold seemed to be singing in the gray air. The waves were ponderous humps, smooth as liquid lead as they rolled and slowly gained momentum, then bursting into fierce frenzies of froth.

The man in the water was a German, the sole survivor of a Ju. 88 which "Duke" Wooley, commander of 124 Squadron, and Tuck had shot down between them while the wing was returning from "Operation Queenie-Orange," a standing patrol off the Dutch and Belgian coasts. He was over forty miles from the nearest land and there were no ships in sight. He had no dinghy. He had no hope.

He was dying slowly, miserably . . . the deadly cold squeezing his lungs flat, cramping and crushing his heart, seeping into his bones and eating into his vitals like an army of savage ants. He might last another ten, fifteen or twenty minutes—choking, threshing, growing numb and weak, so alone and so ridiculous in his helplessness. Every second would be an eternity of despair and torment. Tuck circled low, watching him, and thought *if that were me, down there, I know what I'd be praying for. . . .*

He called up his pilots—"Return to base. I'll stand by here . . . just in case." Just in case what? Could the sea change suddenly to summer warmth, or might a submarine pop up, or a flying-boat swoop in to make the most miraculous landing of all time on those giant breakers. . . ? He knew he was fooling no one.

The wing, which had been orbitting overhead ever since he dropped down to take a look at the place where the enemy machine had gone in, straightened out and disappeared into the darkening haze. He was left alone with the German.

If that were me, down there . . . or if that man were my friend, an old and dear friend, instead of an anonymous enemy, God knows this is what I would want to happen! Yes, I am sure, I am sure. And so I will do it.

He widened his circle, turned in towards the man, heading downwind. He throttled back and put the nose down until the gunsight framed the tiny, bobbing target. He checked the turn-and-bank. His thumb moved on to the little red button. Thank God, from here he couldn't see the man's face, only the top of the head, the bulky collar of the lifejacket and the feebly flailing arms. He pressed the button.

The torrent of shells and bullets lashed the water, making it boil and throw up slender plumes of spray, obscuring the German. Three seconds of that, then the Spitfire flashed over the spot.

The plumes of spray stood up, very straight and tall for a long time before the wind tore them, gradually pulled them aside to reveal—nothing but a wide, white stain which the big waves quickly broke up. Tuck circled a minute or so more, then headed for home.

It was the right thing, the only thing to do. But I will tell no one, for some may not understand.

At the time it did seem right, it did seem the only thing. But later doubt came gnawing at his conscience, a voice that whispered sometimes in the night, whispered a shameful, ugly word. He has asked me to be completely honest about this incident, so I can tell you that today he still hears this voice occasionally, still has to reason with himself to dispel a feeling of guilt.

"All I know is I couldn't bear to fly away and leave

him," he says, "and I couldn't bear just to watch him, either.

"I've never told anyone about this before—though I'm pretty sure my pilots had a shrewd idea what had happened after I sent them on ahead. It hasn't been easy to live with."

How revealing that this one incident should haunt him! Here is a man who in the course of his career shot over thirty aircraft out of the sky, several of them as blazing deathtraps for crews of three or four enemy airmen—one of the many pilots whose guns attacked railway trains, waterfronts, factories and various other targets in occupied Europe, undoubtedly killing and wounding an unknown number of civilians—men, women and, let's face it, probably children. And all these years later, what is the one memory that hurts and worries him? The memory of that single German flier, helpless in the water, whom he shot out of sheer mercy.

Why, then, need Tuck have had misgivings? The answer is that, like many other British fighter pilots, for all his flying days he clung to an antiquated code, and in many senses confused war with sport! As long as the contest was impersonal he was tigerishly aggressive and completely ruthless, but the moment his quarry changed from machine to man, the final whistle blew in his mind. There was, of course, the time at Dunkirk when he'd killed a Luftwaffe pilot on the ground, but then the damned fool had answered Tuck's friendly wave with defiant bullets—it was the German who'd broken the code. . . .

It is a remarkable fact that Tuck, master craftsman and legendary leader, had absolutely no hatred for his foe.

*   *   *

It was wonderful being with Joyce again. After two months away from her, he was more than ever sure that she was the only woman he'd never grow tired of. She wasn't *just* a woman, she was a person, a companion. A pal.

Though marriage was never mentioned again, each

of them understood it was only the war that prevented it. They couldn't even talk about the "one day" in the remote future when, God willing, it would be possible. The future was banned as a topic—they mustn't tempt Fate. . . .

There is a great, destructive sadness that grows slowly in the hearts of young lovers who are afraid to dream. This sadness in time moves them to act very strangely, so that they amaze themselves, frighten themselves, and sometimes wound one another. This happens because dreams are the fabric of love.

There were times now when only the strength and serenity of Joyce's character saved them. Tuck became prone to sudden moods, and drank even more than usual. He was perverse, offhand, domineering, sometimes downright rude. It was almost as if he were trying to goad her into losing that cool poise of hers.

Fortunately she knew enough about him now to realize that the real conflict didn't lie between them, but within him: he was fighting to stay the way he was, the way he'd always been since his first combat —living only for the day; reconciled to death in battle; having nobody but himself to consider in that vital instant when he made a decision in the air.

Joyce by now was one of the very few people who grasped the fundamental truth about his character— that underneath all the flashy assurance, the haughtiness and the forced gaiety, he was a seething mass of doubts and sensibility. It was, in fact, this inner uncertainty that produced his amazing energy, his iron self-discipline, his feverish striving for perfection.

★　　★　　★

Cowboy Blatchford now was commanding the Digby Wing, and they visited each other once or twice. Their talk turned to 257 days, and then came the sadness, the stock-taking. Most of the old mob were gone now —so many names that couldn't be mentioned without a pang, without a slight, awkward pause in the conversation and then an artificial burst of talk that irritated them both, because they were still such close friends and couldn't fool each other for a moment.

Cowboy seemed older, more serious, inclined to be philosophical and at times even cynical. A lot of the fun had gone out of him. It wasn't *quite* like old times any more. The war was changing them both.

Other old pals came visiting too, among them Mike Lister Robinson—the first person Tuck had spoken to at Baker Street Station on that September morning seven years before, as the blazered batch of recruits waited eagerly for the train to Uxbridge. Mike still looked pinkly youthful and a bit lackadaisical—it seemed only yesterday that Tuck had watched him strolling up and down that platform, swinging a tennis racket. . . .

Mike, now a wing commander, had taken over the Tangmere Wing only a few hours before. He was a very experienced pilot, with over a dozen enemy machines to his credit, but for some months he'd been on special duties outside Fighter Command, and therefore felt "a bit rusty." He asked if he might fly with Tuck's outfit for a day or two "and catch up on things," and this was readily agreed. After some convoy patrols, one or two abortive interceptions and a balbo-sweep, Tuck told him: "Get along back to your own chaps, Mike—I can't show you anything you don't know."

Before Mike left, the pair of them put on one of the greatest "beat-up" shows Biggin Hill—that most famous of fighter fields—has ever seen. At first they flew in perfect partnership: the two Spitfires seemed almost like a single, twin-boomed machine, controlled by one mind as, in a tangled skein of noise, they flew upside down inches above the grass . . . grazed the hangar roofs with their props . . . banked gracefully round the control tower. Then they flashed up the main driveway in line astern between the buildings, so low that people watching from the windows looked *down* into the cockpits. They power dived out of the remote blue, pulled out barely clearing the officers' mess, and climbed away like rockets, more or less vertically.

Then they split up, to stunt individually. At once Tuck's superior nerve and sleight of hand were apparent: his aircraft rolled and writhed and swooped and swerved in a continuous, integrated series of maneuvers. After a few minutes Mike was content to circle and just watch, every now and then shouting encouragement over the r/t. Tuck kept it up for almost half an hour—a swift, aluminum needle stitching a mad pattern through the airfield's outer wrappings of huts, hangars, office blocks, telephone lines and hedgerows.

It was a dazzling, dizzying display that is still talked about at Biggin Hill. For those few minutes he seemed to be drunk with power and speed, deliberately, passionately, pushing his skull to the uttermost limits, taking tremendous chances, cutting everything fabulously fine, venturing further and further into the vicinity of disaster. He laughed as he flew, and seemed to glow inside with laughter, to be filled full of primitive joy. Tuck, the incandescent—burning bright, setting the sky on fire. . . . He was proving, to himself as well as everyone else, that he was as good—no, *better*—than ever.

It's easy, he thought. *The right machine, the right procedure clear in your mind, the right amount of care and the right amount of recklessness . . . it's all really very simple. . . .*

Robinson, watching from above, wondered how it was that this one man was allowed to achieve the impossible—allowed to cheat, to throw away the rule book, to sneak by while the Law of Gravity wasn't looking. And when at last Tuck pulled away, wing-waggling down the long, narrow valley that ran along one side of the airfield, Mike called up and said: "Tommy, you're definitely, absolutely, hopelessly off your rocker!" In reply there came a high, goony laugh. "I'm off now," Mike said. "Thanks for the refresher course."

"Cheers, old boy. Keep it in your pants!"

* * *

Biggin Hill's station commander was Group Captain Philip ("Dicky") Barwell, another of those obstinate veterans who refused to stay "chairborne." At this time he was encased from head to hips in a thick plaster

cast, having injured his spine crash-landing a Spit after
its engine had cut on take-off. The plaster didn't deter
him from flying with the wing from time to time—but
it certainly worried those pilots who had to stay near
him in the formation, because Dicky couldn't turn his
head to look back or to the sides, and consequently
was apt to get alarmingly close to his neighbors. Tuck
christened him "Dicey Dicky, the original flying brick."
He admired Barwell very much, and made no effort to
dissuade him from taking part in some of the sweeps
and patrols.

At Brasted, not far from Biggin Hill, stands *The
White Hart,* famous rendezvous of fighter pilots during
and since the Battle of Britain. The landlord, Teddy
Preston, and his wife Kath, got each pilot to sign his
name in chalk on a big blackboard on the wall of the
saloon bar. (The blackboard is still there today, framed
in oak and covered with plate glass. It bears the auto-
graph of almost every British "ace".)

In that December and January, Tuck spent many
an evening at the *Hart,* and when Joyce came for the
weekend she usually stayed there. There was a fine
dining room and the Prestons wrought miracles with
their wartime food quota.

When you went to the *Hart* you were always sure of
meeting up with chaps you hadn't seen in months, maybe years—and to a large extent that still holds good
for air force men. Tuck renewed acquaintance with
Victor Beamish, Al Deere, Max Aitken, Brian Kingcome, Johnnie Kent, Tony Bartley, Peter Hillwood,
Bobbie Holland, Allan Wright, Roy Mottram and many
others. And there, too, he met some of the "up-and-
coming" youngsters like Johnnie Johnson (eventually
to become Britain's top-scoring fighter pilot of World
War II, with 38 confirmed victories), and the quiet,
dreamy-eyed Irishman Paddy Finucane, whose deadly
marksmanship had earned him quickfire promotion to
leadership of the Kenley Wing.

Tuck took an immediate strong liking to Finucane,
but found it very hard to get Paddy talking—he was
extremely shy, and rather stiff and cautious in the company of older, more experienced, or higher ranking

officers. There were persistent rumors, however, that there were very rare occasions when, out with his own pilots, he cut loose and got "screechers"—then, so they said, he was "a holy terror," wild and incoherent and as ready for a rough-house as any God-fearing Cork ditch digger at a wedding or a wake. Once he was supposed to have dived headlong down a twelve-foot basement area in Maidstone, knocking himself out and laying open his scalp. Another report had him climbing halfway up a church steeple in Croydon, going to sleep on a narrow ledge until rescued by firemen with an escape ladder.

Tuck, intrigued by such stories, once or twice deliberately set out to get Paddy plastered, but the Irishman stubbornly stuck to shandies and stayed in his shell. Only a very few people ever got really close to Finucane, and Tuck wasn't one of them. This fact he still regrets—"though he puzzled me very much, I admired Paddy immensely and wished I could have his friendship. Perhaps if we'd had more time together. . . ."

Despite bad weather there was plenty of activity, but few encounters brought decisive results. A few "Rhubarbs," and a lot of fumbling and wild-goose-chasing up and down the coast after intruders in poorest visibility. Only when there was an odd clear spell did they get a chance to go rampaging eastwards on a balbo.

One evening Tuck and two of his pilots went up after a lone raider nosing around a small convoy off the east coast. It was a filthy night of driving rain, gusty wind and low, solid cloud. Nearing the area they split up to search at slightly different heights. Tuck took the "bottom flat" and in an effort to pick up the convoy eased down through the blackness to about a hundred feet. He couldn't even see the surface of the gale-lashed water—not even the brief phosphorescent glimmer of a "whitecap" let alone the dull-glowing feather of a ship's wake. . . .

Several times control called excitedly that one or other of them was very close to the "bandit," but none of them spotted an exhaust glow. After forty frustrating minutes Tuck was getting ready to call it off when all at once a shaft of palest moonlight pierced the murk and, like a weak spotlight, showed him the bomber ahead and to the right, very low over the water.

No sooner did he see it than it passed through the gauzy beam and was swallowed in the darkness again. He swung over and belted after it, dropping to about fifty feet. But after four minutes—nothing. For a moment he thought of going lower—there was always a slightly better chance of spotting an aircraft from underneath at night, because any light there was came from above and outlined it. But he decided against the idea: if his altimeter happened to be a few feet out he'd fly into a heaving wave.

He glanced at the fuel gauge, eased his throttle back and was about to order a return to base when suddenly translucent tramlines of tracer ran out through the darkness ahead of him. It took him a moment or two to realize that the tracer couldn't be intended for him, that the machine firing it must be more or less *under* him! He pulled up a bit, jinked to one side, and peering down over his shoulder barely distinguished the thick-set bulk of a Beaufighter coming up, full speed, on a parallel course, not more than twenty feet above the rollers.

The tracer kept streaming out from it for seconds on end, and then up ahead Tuck saw a red wound open in the night. The wound grew into a knot of flame, and he knew it could only be the German—but he still couldn't discern the vaguest outline of tailplane or cabin or wings! It took him all his time to keep the shadowy shape of the Beaufighter in view, only forty yards off to port. . . .

All he ever saw of that Hun was the momentary outline of its tail unit in the blinding flash it made as it exploded into the sea.

How the hell could the Beaufighter have spotted the bomber in this inky gloom? He couldn't shoot by radar—that only brought him within reasonable proximity, then he had to make visual contact and line up his sight in the usual way. To fly at only twenty feet above a raging, leaping sea on a night like this was an incredible feat. He must have opened fire from five or six hundred yards, and with fine accuracy! That human eyes could penetrate such a distance in almost total darkness seemed like——like Black Magic!

Back at Biggin Hill he described the experience, expressing his wonder and admiration in head-shaking oaths. "That bloody Beau pilot," he said finally, "that bastard *really* must have cat's eyes!"

Many people would have said he was dead-right there. Later he learned that the low-flying Beaufighter had been flown by John "Cat's Eyes" Cunningham, the pioneer of airborne-radar, now the top night-fighter "ace" with an official score of at least sixteen.

Cunningham: earnest little boy with white-blond hair and quiet, shy voice, the imperturbable "Baby-Face . . . !"

On the morning of January 28th a buzz spread around the station that advance units of the American squadrons had arrived in England. Tuck hoped it was true, and that soon he'd be cracking that bottle of Scotch with some of the lads he'd met at Washington, Spartanburg and Wright Field. He decided that in the evening he'd phone an officer he knew at Air Ministry and try to get definite information.

He wasn't able to make that call, because this was the day when Cowboy came over from Digby for lunch, and later saw Tuck, and the young Canadian Harley, off on the "Rhubarb" which ended with the leader's Spitfire crash-landing beside the German A.A. batteries on the outskirts of Boulogne. . . .

17

Slowly he raised his hands, leaning back against the fuselage of his crashed Spitfire, and watched the Germans coming at him. The lower part of his face was stiff and sticky with blood from his nose, and he could see with only one eye. He had to breathe through his mouth, and he kept swallowing blood. His head was splitting and he wanted to be sick, but he stayed erect —a tall, slender figure in a beautifully tailored battle-dress, waiting to be lynched by the infuriated Germans whose comrades he had slain just a few seconds before with a a last, defiant blast of his dying fighter's guns.

Ended now the high adventure; here in this muddy field outside Boulogne, at last the enemy would lay him low. A rotten way to die, at the hands of a yelling mob. Why couldn't it have come in the air? His own fault, for firing that last burst!

What a long, hard road it had been, only to end like this. . . .

They grabbed his arms, three or four of them at once, and dragged him towards the wreckage of the gun-truck he'd blasted. They screeched insults, prodded him with rifle muzzles and butts, kicked him with their heavy, dirty boots. They stopped beside the smoking wreckage of the truck and the broken bodies of the crew. His captors pointed and gesticulated, all yelling at once. They kept shoving him forward, to make sure he saw the full horror of the scene.

Yes, first they rub my nose in it, then they beat me to death . . . or string me up from the nearest tree. He braced himself against the agony and the humiliation.

Then gradually he became aware of a change in their attitude. His arms were free, he wasn't being prodded or kicked any more. His dazed mind fumbled with this mystery and took a long time to grasp the astounding fact. The yelling and the screeching had changed—to laughter! And then—marvel of his life! —they were smacking him on the back and one of them was shouting in his ear, over and over again: *"Goot shot, Englander! Goot shot . . . !"*

Peering with his good eye in the direction of several pointing arms, he saw that by an incredible fluke one of the Spitfire's shells had gone right up one of the long, slim barrels of the multiple 20 mm. and exploded inside, splitting it open in curled strips, like a half-peeled banana! "Tuck's Luck" again—not deserting him in those final seconds of flying, still with him as his dying aircraft leveled out inches from the ground. . . .

To get a better view of the curiously distorted barrel, some of them were walking over the mangled remains of their friends, and still roaring with laughter. The Teutonic sense of humor was incomprehensible to Tuck, but he knew it had saved him.

When their mirth subsided, they led him over some fields to the Flakregiment's headquarters, and gave him a cigarette on the way. A middle-aged *stabsfeldwebel* (senior warrant officer) bathed his face, but couldn't staunch the flow of blood from his nose. Then as dusk deepened he was marched under strong guard for about two miles to a village, where he was handed over to the military police. They noted his name, rank and service number, "frisked" him to ensure he was unarmed, but didn't search his pockets. Then he was locked up in a small, dark cell on the upper floor of what he supposed had been the village hall.

He flopped on to the low iron cot with his sodden handkerchief pressed to his nose, and suddenly began to shake violently all over. His teeth chattered, his breathing became jerky and painful, and though it was cold in the cell he sweated copiously. But quite soon a young, kindly faced Wehrmacht doctor came, staunched the bleeding and gave him some pills. After a few minutes the shaking stopped, and he was able to drink a little *ersatz* coffee and eat some bread and cheese which the guard brought.

The cell had a small, high window with thin iron bars. He thought later on he'd try prying the bars loose with a chair or table leg. If he knotted some blankets together he might be able to climb down to the courtyard.

He rested for about an hour, stretched on his back in the dimness, wondering what was happening right now back at Biggin Hill. Young Harley would have told them all the story—they'd probably think he'd had the chop, that he must have pranged trying to bellyland in that cluttered valley outside Boulogne. He hoped the chaps would remember to give poor old Shuffles his dinner. He peered at his watch: the luminous hands were in a dead straight line. Six o'clock. He could hear the news reader's voice—"one of our pilots is missing. . . ." He wondered if Dicky Barwell had informed his parents yet, and whether somebody had telephoned Joyce.

*   *   *

Barwell had sent a telegram to Captain and Mrs Tuck: "Deeply regret to inform you your son is missing from operational patrol . . . Please accept my profound sympathy letter follows will keep you informed any further news."

They took it very calmly. In a way they'd always expected it. Captain Tuck kept telling his wife and Peggy: "He was missing once before, and he got back. Robert's always had so much luck—why should it *all* desert him at once? Let's keep our heads, my dears, probably there'll be good news in the morning. . . ."

One of the Biggin Hill pilots rang Cowboy Blatchford at Digby, and Cowboy got through to Joyce at Coltishall.

"Listen, honey—I know Bob, and I reckon he got down in one piece *some*where."

"But supposing he's in a dinghy, Peter? On a night like this!"

"If there's one thing I'm certain of, it's that he isn't in a dinghy. He hates the sea—he's always distrusted it, you must know that. In this weather that's the last thing he'd try. Besides, Harley's quite sure he was too low to make it out to the coast. On the level, Joyce, my bet is he's okay—either he's a prisoner, or he's hiding out in some French cellar."

Greatly comforted, she threw some clothes into a case and caught the first train for Walton-on-Thames. She was with his parents next day when a letter arrived from Barwell: "May I tell you how extremely sorry everyone is . . . what a great loss it is to us . . . I am sure there is no need for me to write at length to tell you what a wonderful leader he has been and what excellent work he was doing at this station in so many ways, both in the air and on the ground. . . . There is a good chance of his being safe. . . ."

The morning papers headlined the story. That word "missing" looked venomous in the heavy black type. The *Daily Express* devoted almost a whole page to an article by David Newton, "Tuck—The Man They Couldn't Keep Down."

It was by far the most accurate and skillful piece yet written about him, and Joyce thought it sad that Fleet Street should wait until now to get the facts straight.

In assessing Tuck's character and achievements, Mr. Newton said: "Many of the legends of this war have attached themselves to him . . . he was an 'ace,' an individualist, the greatest individualist fighter pilot the R.A.F. has.

"He has all the qualities, but he has something more. When you have counted the skill and daring, the cool courage which presses on with deadly calculation when terror is just round the corner, you still have to add something to get the full measure of Tuck. It is best described as pride in his work.

"It sounds like a description of a craftsman and not a fighter pilot, where blood runs hot on the job and hatred blurs the vision. But that's how it is with Tuck. He has a pride in the job skillfully done."

✳   ✳   ✳

Before he was taken away on the long trip to prison camp in Germany, Tuck was entertained to dinner by the pilots of the Luftwaffe fighter squadrons based at St. Omer (1) and commanded by one of Germany's greatest airmen, *Oberstleutnant* (Lieutenant-Colonel) Adolf Galland (later General commanding the entire German fighter forces). He was a compact, wide-chested man, dark and of medium height. He had a flat, broken nose, and war and weather had bitten deep lines into his broad face. Around his eyes and on one cheek were small, glossy-pink patches—burn scars. He wore an impressive array of decorations.

"We have met before, *Herr Oberstleutnant* Tuck," said Galland, shaking hands firmly. "Last time I very nearly killed you, but you saw me coming and got out of the way in the nick of time." Tuck remembered the Duxford balbo-sweep on which the two Messerschmitts had come right through from the back of the British formation, holding their fire until they got up near the leader. "So that was you, was it? I got your number-two as he passed in front."

"And I got yours, which makes us—how do you say it—even Stevens?" They both laughed, and then the German led him over to a large, round table set in a bay window. "Come and meet some of my pilots, they have heard all about you of course. Some months ago we had the honor to entertain one of your comrades, *Herr Oberstleutnant* Bader. Wonderful fellow!" As they approached, the eight or nine officer and N.C.O. pilots seated around the table sprang to their feet. Each man as he was introduced to the Englishman clicked his heels and gave a quick little nod of the head, then shook hands. As they all sat down Galland beckoned to a steward, who placed before Tuck a bottle of *White Label* Scotch and a twenty-packet of *Gold Flake* —"about the last of the stock which your Army so kindly left us at Dunkirk," a captain explained.

They talked about drinking, and in particular about French wines; then about food, and the bad weather which had been keeping them on the ground so much. They were careful to avoid any questions about the war or technicalities which might embarrass their guest, but they asked after a number of British aces like Malan and Finucane as if they were old chums temporarily absent. . . .

Tuck was astounded by the feeling of comradeship between himself and these men. Here, in these old-young faces, he saw that same odd mixture of earnestness and gaiety he'd seen so often in other faces at home—yes, and in the United States too. These German faces—he knew them well! Somewhere in this room, somewhere among this group of the enemy, he felt certain there must be a Cowboy, a Jarvis, a Titch Havercroft—yes, that smiling rumple-headed little chap over by the fireplace!—a Mottram, a Kingcome, maybe even a Caesar Hull. . . . What was it he'd thought to himself that time back at Spartanburg when he was briefing the Yank squadrons? *Pilots are a world-wide fraternity: I suppose they look the same, and think and behave the same, even in Germany. . . .*

Galland's last words to him, before he was handed over to his guards to be marched back to his cell, were: "I am very glad that you are not badly hurt, and that now you will not have to risk your life any more. For you the war is over. I hope when it is over for all of us, we may meet again." (They did—in late 1945 Tuck was chosen to interrogate General Galland after he was brought to England as a prisoner of war! They are now great friends.)

During the last hour or so of the evening Tuck had been feeling very sick—he was still woozy from the clout in the face he'd suffered in the crash-landing, and the mixture of whisky and wines and rather greasy food was too much for him. But he'd fought the nausea, because he thought it would be a sign of weakness to let the Germans see his distress. Back at the village hall he spent most of the night vomiting, feeling the spirit running down his nose and stinging the raw flesh inside. He was much too miserable and shaky to attempt his plan of prying the window bars loose.

Before dawn a strong guard party, commanded by a tall young lieutenant with rimless spectacles, came to collect him. The lieutenant saluted smartly. "We are taking you by train to Germany," he said. "You are regarded as an important prisoner and my orders are very clear. If you try to escape you are to be shot down without hesitation. My men will not take the slightest chance with you. Is that understood please, *Herr Oberstleutnant?*"

"Perfectly, thank you very much," said Tuck, and he managed a cheerful grin.

The young lieutenant was extremely efficient. Throughout the entire journey Tuck was guarded day and night by two soldiers at a time, one of them sitting opposite him with a drawn pistol, the other outside in the corridor with a rifle. But there was no bullying: on the contrary when the captive, clad only in his light battledress and still weak from loss of blood and shock, began to shiver and sneeze in the unheated compartment, the lieutenant gave him his own greatcoat. (He took it back, however, whenever they stopped at a station—in case some S.S. or Gestapo man should come nosing around!) They brought him hot drinks, soup and plenty of bread and potatoes. When they reached Halle, near Leipzig, they transferred to a truck and started across country for Dulag Luft— the reception camp for allied air force prisoners.

He was feeling quite fit again. He thought about what Galland had said—about not having to risk his life any more. It was quite true: he need never spend another day in the wartime sky—that jungle of snares and treacheries, where a thousand perils lurked . . . where he'd had such good hunting this far, and got away with it.

Certainly on the record he was the luckiest fighter pilot of all—and now, he supposed, this was the final stroke of Fortune, to be spared at the end, to become a prisoner instead of a charred or pulped corpse. The sensible thing would be to accept the situation, to make the best of captivity. To learn the meaning of patience. He should be content.

But Tuck was not. He belonged to a world whose daily objectives were deeds and not words, action instead of thought, the purposeful blow rather than the acquisition of knowledge. A world in which contemplation was considered a dangerous vice, and sheer survival took the place of self-improvement. An eccentric, upside-down world that coveted and saved that which most men desired to lose, and threw away those things which the conventional majority prized greatly! How could he suddenly change? How could he, after all the wild freedom, all the daring and the doing and the hot thrills, learn now to live in a state of waiting—in a confined space, ruled by a set routine, comforted only by hopes and dreams? His heart grew cold at the thought that the war might drag on for many years—he would rather be dead than drift, derelict and defeated, on a glassy sea of Monotony!

There was something else that influenced his decision—his high sense of duty. He knew the military code demanded that a prisoner of war should try his utmost to escape and rejoin his own unit.

As the truck rumbled through the gates, as he looked over the tailboard and saw the great bastions of barbed wire and the machine gun towers on their high stilts, he decided that he was going to be about the most troublesome, obdurate and insulting prisoner the Huns had ever had to cope with. And cunning, too—he'd get out, somehow! He'd get out, and once more fly for his life. . . .

* * *

On February 2nd, 1942, the German radio gloatingly announced that Wing Commander Stanford Tuck had been shot down and taken prisoner. His parents and Joyce, who had tuned in to every Nazi newscast since Robert was first reported missing, danced a jig round the sitting room. There was laughter, there were tears; Captain Tuck produced a bottle of cherry brandy and they all got a little tipsy. Newspapermen, friends and relatives telephoned, and neighbors came to the door to make sure they'd heard the broadcast.

A telegram from Air Ministry solemnly warned them not to count on Tuck's safety until official confirmation had been received through the Swiss Red Cross, but they couldn't take this seriously—there seemed no reason why the Germans should lie. And sure enough within a few days confirmation arrived.

On the train back to Coltishall Joyce wrote her first letter to him. The first of hundreds. Throughout the years of his captivity she wrote, on the average, twice a week. There were periods of four months and more when she didn't receive so much as a post card in reply, but she never slackened. There was a clear, underlying purpose in this.

Joyce realized only too well how Robert would react to captivity. That ebullient spirit, that boundless energy and haughty exterior were bound to lead him into trouble. Between the lines of each letter she hoped he could read her plea—*don't do anything foolish, darling, not now!*

She knew he'd go on fighting, that his one idea would be to escape. What frightened her was the thought that his natural restlessness might tempt him to do something on the spur of the moment—to embark on some harebrained scheme, instead of waiting for the right opportunity. Her letters—unemotional, full of down-to-earth news about their families and friends, and frequently enhanced by remarkably well-written descriptions of places she visited and interesting characters she met—were intended to have a steadying effect, to remind him, without ever saying it in so many words, that she was waiting, and would go on waiting, cool and unruffled as ever, for five or ten years if need be.

An ocean of ink had been expended—and most of it worthily so—on recounting the experiences of British servicemen in German prisoner of war camps. A large number of books have been written by escapees, and I dare say many more are yet to be published. The story of how R.A.F. pilots in Stalag Luft III, the big air force camp at Sagan, Silesia, continued to defy and, with superb ingenuity and organization, to outwit the enemy, is brilliantly told in *The Great Escape,* by Paul Brickhill. And Brickhill should know, for he was there—in fact, for a time he acted as a messenger for the escape committee, "X Organization," of which Bob Tuck was a member.

To give a full account of Tuck's prison camp life

would not only occupy another volume, but would mean going over much of the ground so fully covered by Brickhill. Therefore, while by no means intending to detract from the heroism and courage of so many other officers who were active tunnelers and escape-workers in Stalag Luft III, it is necessary to rigorously compress the story of the camp as a whole and to concentrate on Tuck's personal experiences, most of which quite naturally are not recorded in Brickhill's book.

Tuck's "private war" began long before he reached Stalag Luft III. At Dulag Luft he was uncooperative and insulting. Neither solitary confinement in a cell that was kept at near suffocation temperature, nor tempting promises of special food and privileges, changed his attitude. He used the foulest language to Luftwaffe intelligence chiefs trying to interrogate him, smashed up his cell after they'd taken away his Polish wings and Iron Cross, and kept demanding all sorts of fantastic things as "rights under the terms of the Geneva Convention."

But his greatest crime, the one which sent the Germans into purple, screeching rage, was stopping another prisoner from giving information.

One day, lying naked on his bed in the "hot box" sweating it out, he heard shouts in the adjoining cell, which he knew was occupied by an American pilot—because the dividing wall was thin, and once or twice before he'd heard an unmistakable, "Deep South" voice calling for the guard. Now the voice had a distinct sob in it, as its owner yelled: "Leave me alo-an, cain't you. . . . God hailp me, Ah cain't stand no mo-ah this!" After days—maybe weeks—of the "heat treatment," and incessant questioning, the American was about to crack up. Tuck pressed his ear to the bricks and heard the interrogators, sensing victory, stepping up their bombardment. Suddenly the American let go a hysterical wail, and then:

"Goddam you, Ah'll tail yo' all you want . . . Ah sweah . . . onl-ay you *gotta lait me outa heah!*"

Tuck leapt from his bed, snatched up a shoe and hammered frantically on the wall.

"For Christ's sake, man! Pull yourself together!" he bawled at the top of his lungs. "You'll tell these bastards nothing—d'you hear?"

For several seconds stony silence. And then a kind of spluttering and scuffling. He heard the Germans burst out of the American's cell and come running down the corridor. Quickly he dropped back on to his bunk, folded his hands across his chest and assumed an expression of saintly benevolence. The door exploded in and a gaggle of semi-apoplectic Huns loomed over him. He let them screech and shake their fists, and all he did was wiggle his bare toes and smile at the ceiling. How easy it was to goad a German into losing his dignity!

When finally they regained some semblance of composure, a fat major boomed in English: "Disgusting—a senior officer ought to behave like a gentleman! You have no right to interfere with us when we are trying to do our duty. This will be reported, and you will be punished. Well—have you nothing to say?"

"Yes," said Tuck mildly, with a parsonical smile. "Bugger off, the lot of you."

Another screeching match, and this time the fat major was literally jumping up and down, thick veins standing out on his neck and brow. It was really very funny, and Tuck was still whooping with laughter long after they'd stormed out and slammed the heavy door. To his deep joy, after a few minutes from the adjoining cell there came a low rumbling, which quickly swelled to a full-throated guffaw. . . .

The interrogators had the American moved at once, of course, away from the "bad influence." But not before he had time to shout: "Ah sho' wanna thaink you, whoevah you are! An' don't worry, Ah'm jus' fine now!" (That Southerner's identity remains a mystery, for Tuck never saw him, or heard of him again. He was probably an American serving in the Royal Air Force, either on one of the Eagle squadrons or with Bomber Command, but it is possible that he was one of the first members of the United States Army Air Corps to fall into enemy hands.)

For this "interference" Tuck was sent along with

about twenty-five other R.A.F. "recalcitrants," to Spangenberg Castle, a medieval fortress perched on a giant pinnacle of rock sticking up unexpectedly in the middle of flat farmland twenty miles south of Cassel in Westphalia. He remembers it as "an enormous old pile, with turrets and battlements and walls about two feet thick—real 'Grimm's Fairy Tales' stuff. It had a dry moat around it, in which the Germans kept wild boars —in a state of semi-starvation. It was supposed to be 'escape proof,' and we were inclined to believe that. But it didn't stop us working out plans."

Yet Spangenberg wasn't nearly as grim as it looked. The majority of the prisoners were elderly Army officers of senior rank, in poor physical shape and too dispirited to think about causing any trouble, and the German staff had an easy time of it.

"Some of the old Colonels and Brigadiers seemed to resent the appearance of air force chaps—we were the first flying types sent there, and probably they were afraid we'd stir up trouble and cause them to lose some of their silly little privileges.

"Mind you, there were notable exceptions—like General Victor Fortune, for instance, who'd been captured after the 51st Highland Division's wonderful stand at St. Valery. The General, and a very famous tunneler from the Kaiser's war named Jack Poole, did all they could to help us, and provided us with a lot of information about the geography of the place, the drainage system and the positions and movements of the guards."

Tuck and Wing Commander "Digger" Larkin, a knife-nosed Australian bomber pilot, decided that the first step should be to dispose of the wild boars in the moat. They inserted rusty nails, razor blades and bits of glass in rotten potatoes and over several days dropped scores of them into the moat: the ravenous beasts gobbled every one—and appeared to thrive on the diet!

Temporarily stumped for a getaway plan, they concentrated in making a great nuisance of themselves. A "go slow" policy, some simple booby traps for tripping up sentries, and sudden uproars in the middle of the night soon had the Germans screeching and spluttering in rage. This "goon-baiting" program was successful beyond their most optimistic dreams: after a few weeks about half of them were "expelled" from Spangenberg—one of the garrison officers actually telling them: "We don't want your type here, this was a *peaceful* place before you came!"

As they marched down the steep, narrow road, they burst into song—a rude version of "Hi-ho, Hi-ho, it's off to work we go," with V-sign actions. The guards suddenly went berserk, and laid into them with rifle butts and even bayonets. There were cracked heads, bruised backs and shoulders, slashed hands and arms —yet when it was over, and they continued the march in silence, there wasn't one of them who couldn't manage to grin.

When Tuck arrived in the sandy compound at Stalag Luft III, among the first people to greet him were two of the original members of 92 Squadron who'd been with him in his very first action over Dunkirk—Peter Cazenove and John Gillies. They looked thinner and older, but they seemed full of spirit and healthy enough. The human race, he was beginning to realize, was insanely flexible.

Then up came Norman "Green to Black" Ryder, "Cherub" Cornish, a dozen others. . . . Here dwelt what remained of a dynasty of heroes.

"Welcome to sunny Sagan! We've been expecting you."

"How'd you like the dragon's castle? Did the Pongos chuck you out?"

"Wotcher, Wings! Don't you wish you were in *Shepherd's?*"

"You should have stayed in America. No movie stars here. . . ."

They knew all about him—obviously there was some mysterious telegraph system between camps. . . .

More familiar faces appearing all around, faces leaping at him out of the hectic past. Boys from other squadrons whose names he'd half-forgotten; others whose names he remembered immediately when he heard them, but whose faces had changed startlingly. Friendly grins, husky handshakes—in a way it was almost like coming home!

He asked about Roger Bushell and learned that his former C.O. was at present in a civilian jail, being grilled by the Gestapo about his latest escape bid. It was very apparent that Roger was a legend with these lads, and there was a fine justice in that. The commander who'd been shot down before he'd had a chance to prove his ability as a leader of pilots in battle had found a kind of greatness now upon the ground: though outranked by many of the newer prisoners, he was the undisputed boss and the constant inspiration of every R.A.F. man behind the German wire!

"He's been out a few times," Cazenove said. "The Gestapo's had several goes at him, but Roger's terrifically tough. What's more, he's a barrister—he answers their questions with some of his own, and finishes up grilling *them!* Not to worry, he'll be back. For a while, anyway. . . ."

Douglas Bader, he learned, had made at least two escape bids. On recapture he'd been so rebellious and insulting that the Germans had decided he was "unfit to associate with other prisoners," and sent him to the punishment cells at Kolditz Strafelager. Old "Tin Legs" —still snarling, still leading the fight! *If a bloke with no legs can do it. . . .* "Well, bless his heart!" said Tuck—and meant it with all of his own.

The atmosphere of the camp was busy and confident. It took him quite a while to find out all the various escape projects which were in hand—several tunnel syndicates burrowing out in different directions from different huts, the tailor shop run by Tommy Guest that converted uniforms into working men's suits, the forgery bureau supervised by Tim Walenn that produced passports and permits, and sundry individual schemes, each of which had to be approved by the "X Organization" to ensure that the escape of one or two men didn't spoil other, more ambitious plans.

Tuck's first escape bid was made in partnership with a tall, rock-faced Polish flight lieutenant named Zbishek

Kustrzynski. They planned to get out in the horse-drawn cart which collected the camp's refuse and took it to a dump on the edge of a pine wood about two miles away. The guards at the main gate invariably prodded deep into the refuse with long, pointed steel shafts, but Tuck decided they could make a shield which would protect them. Once out of the gate and on the way to the dump, it shouldn't be too difficult to unfasten the tailboard, worm out and slip into the woods.

Other officers made friends with the old man in charge of the cart, bribing him with cigarettes and morsels of chocolate from their Red Cross parcels. Gradually they won his confidence, until it became the regular thing for him to slip into one of the huts at midmorning for a ten-minute chat and a smoke. They were careful to entertain him in a hut hidden by others from the nearest "goon-boxes." While the cart was left unattended Tuck and Zbishek were able to take measurements, from which they fashioned a false bottom out of bedboards reinforced with strips cut from a metal basin. They called it "the coffin lid."

They got the all-clear from "X Organization," were supplied with "suits," papers, money and little cakes of concentrated food made from oatmeal, sugar, margarine and powdered chocolate. While the refuse collector was having his break, with the help of other prisoners they unloaded the refuse already collected from other parts of the compound, and lay flat on their stomachs in the bottom of the cart. Then the helpers placed the "coffin lid" on top of them and reloaded, covering everything.

The cart was a little less than half-full. They didn't mind the sour smell, or the dreadful heat, but they'd never have believed that a load made up principally of tins, cardboard, ashes, potato peelings, tea leaves and floor sweepings could weigh so much! During the next hour, as the old man trundled leisurely on his rounds, emptying bin after bin on top of the heap, they found that breathing was becoming hard labor. Crushed against the rough planks, with their heads turned sideways and their sweat-varnished faces inches apart, they cracked jokes in jerky wheezy whispers and stuck it out. But before long the pressure cut off the circulation in their legs, and then suddenly the "coffin lid" groaned and cracked and dust and ashes started to trickle through. Still they stayed put—until Tuck felt an overwhelming drowsiness stealing over his aching body, and realized they were in danger of passing out and being suffocated. "Out!" he gasped.

And found he couldn't move.

Working together, they heaved with their shoulders and hauled themselves towards the back of the cart. They had to fight for every inch. It felt like being caught in the deep, sucking slime of a swamp. Another sickening crack, more dirt poured through. It kept coming, faster, piling up between them and the tailboard. They were being buried alive.

Worse than being lynched, Tuck thought, *to die in a stinking refuse heap. . . .*

They gathered their strength for a final effort. Together they heaved, together they thrust forward. Tuck's outstretched fingers, clawing through the loose dirt, found the catch of the tailboard.

Seconds later the refuse collector rounded the corner of a hut with one of the very last bins on his stout old shoulders in time to see his load cascading over the back of the cart, and two gray phantoms staggering out of the dust, coughing and retching. Other prisoners came to the rescue—rushed Tuck and Zbishek into the hut, and even remembered to retrieve the broken "coffin lid," before any of the guards could come and investigate.

When a guard did appear, the old man said it was just an accident—the tailboard catch had given way. He couldn't afford to admit that he'd been accepting hospitality from the prisoners and leaving his cart unattended.

Next time Tuck came much closer to success. There was no discomfort, either. He simply tried to walk out through the main gates.

From time to time the Germans took a party of prisoners out to be "deloused" at a sanitary center a mile or two distant. "X Organization" planned some unofficial outings, providing their own guards—prisoners dressed in German uniforms made in Guest's tailor shop, and armed with dummy pistols in cardboard holsters which had been rubbed with boot polish until they shone like leather. Walenn's bureau provided beautifully forged passes to be shown at the inner and outer gates.

First, twenty-four men with two stiff-backed *"unteroffiziers"* as escort marched to the main entrance, chatting and laughing as naturally as they could. They carried rolled towels under their arms, ostensibly bound for the delousing showers. Wrapped in their towels were maps, civilian suits, and concentrated food cakes. At the inner gate the guard didn't even bother to look at the pass. At the outer gate, the forgery that Walenn's crew had taken a week to perfect was given only a perfunctory glance—then the party were trudging merrily down the main road. About a quarter of a mile away they rounded a bend, dived into the woods, changed into their traveling clothes and split up into ones and pairs to make their bids for freedom.

Now it was the turn of the second, smaller party. Bob Van Der Stok, a Dutchman who spoke perfect German, was escort in charge of five senior officers supposed to be on their way to "a special conference with the Kommandant at his headquarters:" Tuck, Squadron Leader Bill Jennens, "Nellie Ellan" (who'd looked after the camp's secret radio), an American colonel and a Polish wing commander. They got through the first gate without trouble, but the second guard happened to turn the pass over and looked at the back. Unknown to the prisoners, just six days earlier the Germans had decided to put a new mark on the reverse sides of all gate passes in case they should be copied. . . !

There was a hell of a furor; the entire camp was paraded for roll call; troops with Tommy guns searched the huts and within an hour the hunt was on for the twenty-six missing men. None of them got back to England. One by one they were brought back and dumped in Sagan's "cooler" for two weeks solitary confinement; and this time Tuck and the other members of the second party did join them.

Roger Bushell came back. A very different Bushell from the one Tuck had known back on 92. No longer ready to play the fool, but grimly, quietly earnest. Still ringing in his ears were the screams of people being tortured by the Gestapo in adjoining cells . . . the crackling of the firing squads drowning out the *matins* bells. A Czech family who'd sheltered him in Prague during one of his escapes had been slaughtered without any sort of trial. He'd been told very plainly that if he caused any more trouble he'd go against the wall himself. Deep and cold ran his hatred for the Germans: he was obsessed with the desire to keep on fighting them, to beat them yet.

Immediately he took over as "Big X" and in no time at all he built up a complex organization which simultaneously pushed out three major tunnels—"Tom, Dick and Harry." The tunnelers worked with air conditioning, electric light, and trolleys that ran on rails to carry back the diggings. It was all the outcome of Roger's brainwork, determination, his knack of getting the very best out of other men's talents. His pernicious faith in the impossible.

Tuck shared a room with him. Bushell wanted to know everything that had happened on 92 after he'd been shot down. And when he'd heard details of every combat, all the benders, and the fortunes of the individual pilots, he'd go right back to the beginning and ask: "What happened while you were still at Hornchurch—before you went to Wales?"

"Roger, I've told you all that, a dozen times over."

"Well, tell me again."

And Tuck would—because he knew how proud Roger was of 92's success, and that the greatest sadness of his life was that he hadn't been able to stay leading them at least a few weeks longer, and sample Maloney's "bubbly. . . ."

Things had changed a lot since Hornchurch. Now Tuck was the one with the flying experience, the decorations, the superior rank. Yet Roger was still the boss, every bit as much as on the first day when Tuck had reported to him to take over as a flight commander. Now, as then, they respected one another. Bushell's strength and realism comforted Tuck when his restlessness boiled up. Bushell's cunning delighted him, and made him see the value of careful thought. Tuck's energy and enthusiasm, haughty bearing and steely nerve gladdened and reassured Bushell.

Slowly their friendship ripened. Roger was the last of the three really close friends Tuck has had in his life.

And in those days he stood in great need of a new friend, because one by one the old ones were going. Over the secret radio, or from incoming prisoners, he heard of their deaths—Paddy Finucane, hit by ground fire not many miles from the spot where Tuck himself had been clobbered, and lost trying to ditch in the sea. Victor Beamish, bounced—as Tuck had always feared he would be—while playing the lone wolf . . . young Ronnie Jarvis . . . dear old Dicky Barwell . . . Mike Lister Robinson. . . .

Cowboy.

The sea. The blind, indifferent sea. The stinking black ooze, the unexplored darkness. Down, down into oblivion, the last, silent dive. Cowboy, still in his cockpit, to rot or to burst along with his Spitfire, to join the squadrons of the deep—there must be thousands of them down there, British and German.

He would never look at the sea again without thinking of them all. Down there, blind-flying through Eternity. He would never look over the rail of a ship or out from a promenade without a pang of grief for that chunky, chuckle-voiced Canadian. Nor without anger.

He couldn't forgive the sea. Neither could he forgive so many of his friends for dying. He looked on these vanishing acts as betrayals, and deeply resented them.

And yet, for all that, he couldn't deny a certain perverse thrill, a certain sweetness, in not being dead himself!

Some mornings, when he awoke very early and

Roger was still asleep, in the deep quietness before the start of living, from far away he would think he heard a low, familiar roaring and crackling: the brave old warsong of the Merlins. . . . Then he would lie a long time listening, and in his mind's eye see again the lines of fighters drawn up in front of the dispersal huts, with the ground crews bustling to and fro and the pilots hurrying out with their 'chutes slung over their shoulders and spotted scarves round their necks and shouting and laughing to each other and clambering into their cockpits and revving up the engines and giving thumbs up signs for the chocks to be whipped away and calling over the r/t and the ether crackling and suddenly control breaking in with *scramble, scramble*. . . . Then he would come back to the present, the vile realities of the prison camp—filthy food, ragged clothes, packed compounds, foul and reeking gutters. . . .

These were the days when he wanted so desperately to get out, to fly a Spitfire again. These were the days when he would avoid the others and pace the compound alone, his face mean and wolfish, his eyes flicking from side to side, his nostrils flaring to catch every whiff of wind coming in from the wide, free forestlands beyond the wire. These were the days when Roger worried about him—he could never forget that Tuck was a great individualist. "X Organization" would hurriedly find some job to keep him occupied.

A batch of prisoners arrived back from punishment camps and jails, and stomping along at the head of the ragged column, growling profanities, was an unmistakable figure: Bader. The guards eyed him in awe, falling back under the lash of his tongue. Some automatically clicked their heels and stood to attention as he approached!

While the column was being checked in through the *vorlager,* prisoners in the main compound crowded up to the warning wire, shouting and waving as they recognized friends and squadronmates. Suddenly Bader saw Tuck. Ignoring the guards, he turned out of the column and came lurching right up to the dividing wire at a point close to the gate where there was only a single fence.

"Well, I'll be damned," he bawled. "Old Friar Tuck!" And he stuck his right hand through.

Tuck accepted the challenge. Smiling broadly, he swung a long leg over the warning wire, walked the five forbidden yards to the fence and grasped the outstretched hand. Eight hundred men cheered themselves hoarse while the Germans looked on, silent and helpless.

"In that moment Bader was the leader of us all," Tuck recalls. "I think if he'd ordered us to attack the guards with our bare fists, every man would have obeyed without hesitation. He treated the Huns with tremendous contempt, and we loved him for it.

"Douglas may be an infuriating character at times—utterly unreasonable, obstinate and pretty damned selfish. Nevertheless he's a born leader of men—on top of his great personal courage, he possesses the power to stimulate unquestioning devotion and to strengthen the confidence of the most doubting souls. I never went into battle with Bader, but I saw those qualities demonstrated clearly enough that day at Sagan."

But Bader and Bushell, the two dominating personalities, had strong differences, and usually Tuck had the unenviable role of peacemaker. Bader didn't find it easy to submit to "X Organization's" directions or fall in with the long-term policy. Just when Roger wanted all the guards lulled into a sense of false security, Douglas would suddenly flare up and insult one of them, or deliberately pick a fight with Von Lindeiner himself.

The problem was eventually solved by the Germans. They moved Bader away again. But it took them quite a time, and nearly forty guards with fixed bayonets were needed. He simply stood with his tin legs wide apart, puffed at his pipe and hurled abuse, until at length it seemed that one of the Huns would lose his head and press the trigger or lunge at him. It was then that Tuck shouted: "Douglas, for Christ's sake—don't push them *too* far!" Others took up the appeal, Herbert

Massey reasoned with him, and Bader finally gave in. He stomped off, surrounded by a small army, bawling lewd remarks about Hitler's mother.

Some of "X Organization's" tunnels were discovered and destroyed by the overalled "ferrets" the Germans sent in to probe about under the huts and in among the drains. Others ran into soft sand and had to be abandoned after falls. But "Harry," which had its trapdoor under a heavy stove in Hut 104, remained whole and undetected. Week by week they burrowed on—under the first double fence . . . across beneath the *vorlager* . . . under the roadway and on towards the forest. Almost four hundred feet long, now.

Bushell began his final preparations for a mass break. He checked over his "stores"—more than forty civilian suits; several thousand marks in the "Holiday Fund;" plenty of concentrated food, cardboard suitcases, compasses, railway tickets, passports, identification papers, travel permits, even handwritten German phrase-books. . . .

Bushell never went near the tunnels—because of his record he was always closely watched by the "ferrets" and guards. He began rehearsing in the camp theater for a production of *Pygmalion,* casting himself as Professor Higgins—though he had every reason to hope that on the night of the show, three weeks later, an understudy would take over "in the unavoidable absence of the star." Once, while Von Lindeiner and other German officers sat in the stalls watching a rehearsal, Roger stood in the wings and discussed with Massey, Wings Day, Tuck and others a scheme for drawing lots to decide the order in which the men should go through "Harry" on the big night. He thought with luck anything up to two hundred might get away between sunset and dawn.

They learned about this time that Himmler had sent word to all prison camp Kommandants that sterner countermeasures were to be taken against escapers, and had hinted that from now on the Gestapo would claim recaptured officers as their prisoners, to be dealt with as "saboteurs and spies." But there wasn't one man who wanted to back out.

Then one morning during roll call parade, a strong party of extra guards marched into the compound and took up position between the prisoners and their huts. Nineteen men were called out of the ranks and, without a minute's grace to collect their few poor belongings from their rooms, were marched straight out of the camp under heavy escort.

Tuck was one of them.

They were taken to a smaller camp called Belaria, about six miles away on the other side of the town of Sagan. For hundreds of yards all round this camp the flat ground had been cleared: the little cluster of huts stood in windy isolation. The enclosure bristled with searchlights and guns, and the guards walked about in twos and threes with their fingers hovering close to their triggers.

The nineteen presumed they'd been "purged" because they were suspected of escapist activities. Most had records as "trouble-makers." Zbishek Kustrzynski, Peter Fanshawe, Wally Floody, Jim Tyrie, George Harsh—all cursing, all sick at heart. They'd missed the mass break by a few days, after all the months of toil and tension. . . .

Even so, Tuck chuckled to think that the Germans hadn't taken "Big X" or several other key men! Bushell's brainwave about producing *Pygmalion* had provided a wonderful cover for most of the top organizers. Tuck had been offered a small part in the play, but he'd turned it down.

The move to Belaria proved yet another of those amazing pieces of luck. On a moonless March night not long afterwards, seventy-six Sagan prisoners bellied through "Harry" in what is now known as "the Great Escape." All but three were recaptured.

Fifty, including Roger Bushell, were murdered by the Gestapo.

Evidence which was accumulated after the war, and used at Nuremberg when twenty-one of the killers were tried (twenty were convicted, thirteen hanged) indicates that the fifty victims probably were selected with an eye to their previous records in captivity. None who'd given serious trouble in the past was spared

The Gestapo thugs drove their prisoners in twos and threes into the countryside, pushed them out of their cars and shot them—mostly in the back. Then they cremated the bodies—to hide the exact manner of death—and announced that the officers had been shot "attempting to escape" or "resisting arrest."

By missing the break-out, Tuck had missed death. He would have gone with Roger, and been recaptured with him. With his record—"non-cooperation" and "insulting conduct" at Dulag Luft, a stretch at Spangenberg, attempted escape from Sagan and various other acts of defiance—he certainly would have gone on one of those motor trips. . . .

When the news of the murders reached Belaria, Zbishek and several others had to restrain Tuck, forcibly, in his room or he would have done something suicidal. Only once or twice before in his life had he lost his temper: this time, for two hours or so it really seemed that the balance of his mind had gone. A dreadful exhibition.

At last he collapsed, played out, on his bunk, and lay for a long time without making a sound, oblivious to the others in the room. Not since he was an infant had he shed a tear, but now he had to screw his eyes tight to hold them back. (Today he keeps three photographs on his dressing table: Caesar, Cowboy and Roger. All in black frames.)

Kustrzynski stayed with him all night. Tuck hardly moved but, just as a mud-colored dawn appeared he suddenly said: "Zbishek, you and I are bloody well getting out of here!"

"Right," said the big Pole. "I'm with you."

Tuck as a pilot had never hated his enemy. As a captive, even after the murder of the fifty officers he found it impossible to hate the stolid, simple, soldiers who guarded him, much less—as some of the others now did—stoke up a blind hatred for the whole German race. He confined his fury and loathing to the real Nazis—the maniac Hitler's blood-crazed "supermen," the Gestapo thugs, the ambitious bureaucrats— all the jumped-up, death-drunk dregs that now ran this country.

He could see only one way of hitting back at the Nazis—by fighting on as Roger and the others had done. By escaping, in spite of everything.

But it took a long time. After the Sagan break Von Lindeiner had been arrested, to face court martial for "neglect of duty," and three of the camp's maintenance staff had been sentenced to death and shot as "saboteurs" because they'd failed to report the loss of tools and materials which were stolen by the prisoners and used for tunnelling. Belaria's Kommandant and guards, badly shaken, redoubled their vigilance; "ferrets" prowled ceaselessly in the camp; the open ground all round was floodlit after dark except during air-raid alarms, and all prisoners' privileges were withdrawn.

Weeks, months, went by and one plan after another had to be abandoned. Often Tuck's patience snapped, but two things sustained him, prevented him from embarking on some wild and futile enterprise—the swelling roar of the Allied bomber streams thrusting for the heart of the Fatherland, and Joyce's sane and steady-flowing letters.

In the first days of 1945 the food situation suddenly worsened. It had never been good—the Germans, in defiance of the Geneva Convention, fed each prisoner on a budget of roughly one shilling and twopence a week, and but for the Red Cross parcels, thousands undoubtedly would have starved. Now, as the bombing of railways and cities increased and the victorious Russian armies pressed towards the Polish border, the parcels stopped coming through regularly. Tuck, always fond of food, suffered dreadful gnawing pains and stomach cramps, and grew frighteningly thin.

At six o'clock on January 28th (that fateful date for Tuck!) the Germans suddenly routed them out and marched them off through thick snow, heading due west. The thinly clad, ill-nourished prisoners suffered frostbite, developed coughs and colds—some of which led to pneumonia. Roland Beaumont (later a famous jet test pilot) lost his voice and probably had a touch of pleurisy. Many collapsed and had to be carried by their mates. A few became delirious, shouting incoherently, above the howl of the wind. They were prodded and pushed along, hour after hour, and no amount of protest by Tuck and the other senior officers won any respite. The guards seemed very nervous and quick-tempered.

Darkness had fallen when at last they halted at a farm near Kunou, Upper Silesia, and were herded into barns and outhouses. That night men cried out in their sleep. Two fell silent forever.

Next day they stumbled on, through the unrelenting snowstorm, to another village called Gross Selten. By this time there were so many sick that even the most panicky guards took pity, and they were allowed to rest for two nights. But they got no medical attention, and the only food was watery stew. On the 31st another desperate march brought them to a place called Bransdorf, where they were once more locked in barns. They were given a very little thin soup. Somebody produced a packet of pepper—saved from a Red Cross parcel. They poured it into the soup: the sting of it on their tongues gave an illusion of warmth.

Tuck's nose, lips and one ear were severely frostbitten. Zbishek had diarrhea. Almost every man had some ailment—another day like this and many of them would die. They still had no inkling where they were being taken, and the notion germinated that Hitler might have ordered the execution of all Allied prisoners before the advancing armies could save them. Or maybe the Gestapo wanted to liquidate all those who knew something about the Sagan murders.

It was generally agreed that now it must be "every man for himself." Anyone who saw a chance to "duck out" should take it.

Tuck and Zbishek decided to stick together. They explored their barn and found that its loft was stuffed with straw, twenty feet thick towards the back. While they were trying to formulate a plan, the guards let in two teen-aged boys to collect the empty soup pails. They didn't look German. Zbishek addressed the elder of them, a lad about sixteen, in Polish and learned that they were Russian slave-laborers.

"Will you help us?" Tuck asked in suddenly remembered Russian. The boys looked scared, but they nodded in affirmation. With Zbishek's help he explained that they meant to hide under the straw in the hope that they'd be left behind when the column moved off next morning.

"If you succeed," the older boy said, "we will try to bring you food." Then he hurried off, pulling the younger one after him.

They said goodbye to the others—"just in case we do get away!" Roland Beaumont, inexplicably, seemed a lot better than on the first day, but he was worn-out with coughing, still had a streaming cold and was very depressed. "Robert, I don't think I'm going to make this. If you get home, go and see my wife . . ." Tuck cut him short. "You'll probably be back before us, Rolly!" Beaumont cheered up, shook hands and said: "See you in London."

Zbishek arranged with some of the chaps for a "cover-up" at roll call in the morning. By now they were all past-masters in the art of slipping from one rank to another while the Germans were counting, thereby making the addition come out the way they wanted it to.

Tuck found the pepper packet, still half full, and put it in his pocket. Then he and Zbishek clambered into the loft and burrowed deep into the center of the straw. It was soft, loose and not heavy. Tuck pushed his way upwards for about ten or twenty feet and holding his breath carefully sprinkled the pepper over the area above their hiding place. Then he wormed his way back down and said: "All set, chum. Well, at least this is better than the refuse cart!"

"I wish us the best of luck," the Pole said solemnly.

It was warm under the straw. Huddled together, they fell into a deep sleep.

18

About five hours after they'd heard the others moving off, the search party came crashing into the barn. The N.C.O. in charge thundered at his men to find pitch-forks and probe every square foot of straw. "Wish we'd brought the coffin lid," Zbishek whispered. Tuck silenced him with a jab in the ribs.

Then they heard the dogs barking. . . .

The German ransacked the lower part of the barn —overturning machinery, smashing the wooden stalls —then their boots came thumping up the broad stair-boards to the loft. Several of them climbed on top of the straw. The fugitives felt the pressure as each man passed above them. The *hundfuehrer* sounded very close as he shouted commands to his Alsatians——

"Hans! Karl! Good boys, smell them out!"

The rustling movements of the dogs over the straw grew louder, stopped directly overhead. The sound of their heavy panting carried down very clearly through the stack.

"Over here, all you men!" the *hundfuehrer* cried suddenly. "Let's shift some of this!"

Pitchforks swishing and tearing through the straw . . . every stroke a little louder, a little nearer . . . the Alsatians whining in excitement now—probably being held back on their leads while the soldiers worked.

"That'll do, stand back and let the dogs in! Smell them out boys, smell them out!"

The dogs came scampering, pawing, sniffing. To Tuck and Zbishek, it sounded deafening, as if they were only two or three feet above. Then the quick breathing changed abruptly—to sneezing.

The Alsatians didn't like the pepper. They got out of the straw at top speed and bounded away to the other side of the loft. The *hundfuehrer* cursed and went crashing over the straw after them, bawling threats.

Tuck and Zbishek waited for the digging to be resumed, but after a few more random prods the soldiers moved away and worked on another part of the stack.

"Ach, these stupid dogs," they heard one man complain, "they think we're playing a game, that's all. They'd make us move tons of this stuff, just to get at a dead rat. . . ."

The *hundfuehrer*, judging by his furious shouts, also subscribed to this theory. At any rate, he didn't bring the dogs back. After another quarter of an hour, the whole party left.

The fugitives stayed in the straw for the rest of that day. They were ravenously hungry, they were excited by their success and eager to start on their way, but they knew they must move only by dead of night. They had decided to make for the Russian lines—surely not more than a hundred miles to the east. . . .

A low, persistent whistling brought them out of the stack. The elder of the two Russian boys was waiting, holding a cloth bundle in his arms. He was a melancholy, soft-eyed, soft-spoken youngster, with shaven head and narrow chest. As he unwrapped his bundle he told them his name was Shenia Sukovkin. They gave him their names, and thanked him for taking risks on their behalf.

He had brought them half a loaf of black bread and some potatoes. Also some very ragged clothing, thick socks, warm ski-caps with earflaps and a long-bladed hunting knife. When they'd gobbled the food and donned the caps and oddments of clothing, he said: "Now you are to follow me," and led them out of the barn.

They kept away from the roads. Though the blizzard had ended at last, the going was very rough. Snow lay deep in the fields, and a fretful wind blew it into drifts, whipped it up in their faces. The countryside was bare and still and empty of all wild life. *The time when everything turns to white iron.* . . . Slowly and insidiously the terrible cold crept into their gaunt bodies. Frequently they lay down in ditches and behind hedges, waiting for vehicles to pass along a nearby road. Every sound and glimmer of light had a double meaning, every move a dubious consequence.

Crossing a frozen river, Shenia slipped and fell heavily. Zbishek stooped to help him, but with a flurry of indignation the boy shook free of his grip, and got to his feet unaided. As he set off in the lead once more he held his shoulders well back, but he couldn't help limping just a little.

"He's a good chap," Tuck said, through chattering teeth. "All Ruskies are tough!" Zbishek only grunted.

The hard miles faded behind them, and the thumping aches under their ribs became a torment. Daylight was beginning to seep through the clouds in the east when they staggered into a farmyard not far from Gross Selten. Shenia led them straight into a barn where they were gladdened by the sight of a large, charcoal brazier with a pan full of ersatz coffee simmering on top, and a black cat sitting nearby, polishing its whiskers. As they threw themselves down and basked in the blue-red glow, a small figure emerged from the shadows. Shenia introduced Vasya Beresnev—another Russian boy, a little younger, and with a deep, ragged scar across his forehead. He handed Tuck four thin, evil-looking cigarettes, smiling and bowing.

"This is the English Colonel," Shenia said. Then, waving a hand towards Zbishek, but not looking at him —"and that is Captain Kustrzynski." There was something in the lad's manner that troubled Tuck.

Seated around the brazier, sipping the scalding coffee and munching potatoes, the two pilots and the two urchins held a council of war. The countryside, the boys said, was thick with Germans. The Russian forces, it seemed, were considerably further away than Tuck and Zbishek had thought, but were advancing very swiftly.

"We have many friends on farms between here and

the front," Shenia said, "*Ostlanders* like ourselves. We will pass you on to them. They will hide you and feed you. And when the time is right, Vasya and I would like to join you. We shall all be liberated together!"

Tuck thanked them, and asked whether it would be possible to find a couple of old overcoats. They said gravely they'd see what they could do.

He presumed it was because he held the senior rank that Shenia spoke mostly to him rather than Zbishek. Fortunately his Russian was improving with every sentence, and he could understand the boys perfectly.

When Shenia and Vasya had gone, and they lay down to sleep Zbishek said: "Look, Robert, I think from now on it would be best if I spoke only English."

"Why?"

"Poles and Russians don't mix. Thought you were a history student?"

"But Zbishek, you're an officer of the Royal Air Force. They'll have to treat you as such."

The Pole's craggy face crinkled into a tired smile. "Wouldn't count on it. And when we contact them I can't give a false name. Sooner or later they're bound to check with London, or the British Embassy in Moscow, or something. . . ."

"All right, then, say you're English. Say your *father* was Polish, but he settled in England many years ago, married an English girl and you were born there. If there's any trouble, stick to that story. Nobody can disprove it."

"Right, that's the drill. And from this minute on, I can't speak Russian, all right? That kid picked up my accent the first time I opened my mouth—I'll bet he could tell you what province I come from!"

"But you'll have to use a *little* Russian, that can't be avoided. Try it with an English accent, though—the way I speak it."

"That," said Zbishek, "I could never do!" He rolled over, flicked the end of his cigarette into the brazier and sighed.

"I'd hoped all this would be finished now," he said, "but it was stupid of me—that boy's face showed me that! I've a terrible feeling that Poland isn't going to get much of a victory out of this bloody lot. Now wouldn't that be damned funny? Half the world goes to war over Poland. Now they're knocking the hell out of Germany, and when it's all over they'll all be so pleased with themselves they'll forget what it was they set out to do in the first place!"

* * *

In the next three weeks they moved from farm to farm, working slowly eastward. Always they were looked after solely by Russian slave-laborers, whole peasant families the Nazis had transported from the Ukraine to work on Polish and German farms. These people had very little food for themselves, but they readily shared what they had.

Zbishek was weak and staggering from dysentery, but he remained resolute, and forced on gamely without a word of complaint. Tuck marveled at his determination, wondered where he got such reserves of strength.

Once they jarred awake and found their hiding place surrounded by Germans—bedraggled, surly infantrymen, their faces blurred with fatigue, obviously just back from the fighting zone. Tuck and Zbishek scuttled to the loft of the barn and, peering through a narrow slit at the eaves, watched them plundering the farm of food and all the loot they could carry away.

Then one of the Huns, a plump N.C.O., came into the barn and started climbing the steps to the loft! They dived for the only possible hiding place—under the water-tank.

The German walked right up to the tank, lifted the lid and looked inside—for some reason checking the water level. There wasn't much light in the loft, so he propped up the lid and struck a match. As he did so, he shifted round one side of the tank and his heavy boot came down on Tuck's fingers. Tuck had to bite his lip to check a yelp of pain. It was several seconds before the big man shifted his foot again and went back down the steps.

Tuck's hand was stiff and one of the nails started to turn black. But an hour later the bedraggled column trudged off, and the danger was past. And that night Shenia turned up with two worn but windproof coats. When Tuck asked where he'd got them he said calmly: "Off some people killed by strafing." Only then did they notice the bullet holes and the dark stains. . . .

Just a few days later they were at yet another farm when a whole convoy of trucks filled with troops and equipment suddenly turned off the road and came rumbling into the yard. It was plain that the Germans meant to billet themselves here for the night. Two husky young Russians grabbed Tuck and Zbishek and stuffed them into a unused ham-curing oven, a brick box approximately four feet high, three feet wide and five feet long, in the spacious entrance hall of the farmhouse. They couldn't stand up, they couldn't lie down. They could only squat, heads bowed and knees drawn up, in the inky darkness. Within half an hour their muscles were cramped and throbbing and their spines seemed to be cracking. But after a while longer they grew numb, so that their discomfort lessened. Fortunately there were some small holes in the metal door, so they wouldn't suffocate.

They had hoped to slip away after dark, once the foes were asleep. But sentries were mounted in the hallway, and all night long they could hear footsteps, low voices, rifles rattling. Once a sentry stopped by the oven and they heard him strike a match on the top.

For the past few days Zbishek had been suffering from dysentery. Towards morning he had an attack. And then, of course, Tuck was nauseated. Zbishek's high Polish pride was shattered by the situation.

Suddenly they were cursing each other in the blackness. But neither had much strength; and soon the effort exhausted them.

They were silent for a long time, then Tuck whispered, "Zbishek, that was bloody awful! I don't know how it happened."

"Neither do I. Let's forget it."

"How d'you feel now?"

"Exhausted, but I can stick it a bit longer if you can, Robert."

"Good lad. Look, I'll bunch up really tight . . . like

this. Now try moving your legs a bit, eh? Then I'll have a go." In this way they found they could flex their numbed limbs for a few inches. Yet they had to take great care not to kick the door: the slightest sound might have attracted a guard.

But Zbishek had more attacks of dysentery.

"Christ, I'm sorry, Robert! I'm sorry!"

"It's *all right*, boy. . . . These bastards will move soon as it's light."

But the Germans didn't move. The fugitives had to stay cooped up in that stinking black box.

For forty-two hours.

When at last the trucks rolled away and the Ruskies opened the oven door, neither man could move. They had to be lifted out, still doubled up. Necks and spines and legs were locked rigid. They had to be massaged and gradually pulled back into shape, and it was three days before they could walk properly again. But then their spirits soared—they'd stuck it out, they'd cheated the Germans once more! If they could get through that, they'd get through anything. . . .

On again by night, ever eastward. Now in the distance they could see the faint flashes of gunfire. By now they both had thick beards, and in their ragged coats and woolen caps they looked like a couple of anarchist assassins.

Zbishek got over his dysentery, and they began to talk a lot about what they'd do when they got back to England. Tuck knew what he'd do. He'd marry Joyce just as quickly as the law would allow! Joyce, with her enduring serenity, her unspoken devotion, her undiminishing faith. . . . The war would soon be over, and they'd have their future. He'd stay on in the Air Force, of course—he was a regular officer—but one day they'd have a permanent home, somewhere in the country. Kent, maybe. There were so many lovely houses in Kent that he'd gazed down on from his speeding Spitfire. Into his mind visions kept drifting: a big Elizabethan fireplace, a heavy oak dining table, a silver teapot, a front door with wrought iron on it. How he'd value comforts after this!

Meanwhile: yet another barn, another crisis.

A large group of French prisoners were shoved in by their escorts. As soon as the doors had been closed, Tuck and Zbishek crawled out of the hay and introduced themselves. The Froggies seemed only casually interested. They were a volunteer working party, and seemed to be extremely well off for food, cigarettes and clothing. They even had some bottles of beer and wine! But they didn't offer to share anything with the R.A.F. men.

During the night a faint creak wakened Tuck. Half-opening his eyes, in a thin shaft of moonlight slanting down from a skylight, he saw one of the Frenchmen bending over Zbishek and reaching one hand into the tattered cloth bag which held the Pole's few belongings and some scraps of food. Tuck sat up very slowly and drew from his belt the hunting knife Shenia had given him. He balanced it along his palm, the way Portuguese Miguel had taught him so long ago on board the *Marconi*, and threw it with all his strength.

A thud, a shriek, then pandemonium. Cackling, arm-flapping Frenchmen milling round, bumping into each other . . . the would-be thief sobbing and yelling, his hand transfixed and pinned to the wall.

Tuck got up, shouldered his way through the rabble, grasped the man's wrist and with a quick, strong pull extracted the knife. The man shrieked again and slumped down, unconscious. A howl of fury went up from the others. Zbishek jumped to Tuck's side as they closed in. The big Pole's teeth were bared, and he grabbed one Frenchman by the hair and with an easy twist sent him rolling across the floorboards. Now other blades gleamed in the moonlight. It didn't seem to matter to the Froggies that their comrade had been injured in attempting a robbery.

Then through the milling mob came a miniature cyclone—a middle-aged colonel, not much over five feet tall, with a face rotund and red as a paper lantern. He laid about him with a stubby cane, bowling men over on either side. In no time at all he had them all scuttling back to their sleeping places. Then he bowed to Tuck and Zbishek and made a long and profuse apology.

His men, he explained, were completely demoralized. They had all been prisoners since 1940, and they'd forgotten the decencies of civilized life. The Englishmen had been quite right to defend his friend's property.

Tuck was worried about being given away to the German guards, but the colonel assured him that his men knew very well that he, personally, would flay and castrate anyone who even threatened such a thing. "I let them do as they like most of the time," he confessed, "because, frankly, I do not care to wear myself out shouting at them all day, every day. But when I *do* give an order, they know I am in earnest!" This claim was justified: in the morning when the Germans asked why one of the men had his arm in a sling, the colonel said: "Oh, they are always fighting, but I have dealt with them." Not a murmur from the others as they formed into line and shuffled off.

Soon after first light on February 22nd, 1945, they were wakened by the crash of shells landing close to their barn. The flashes of the explosions were searing through the loft, and tiles were flying off the roof. They crawled under the water-tank a moment before the rafters collapsed. Peering out of the wreckage, they saw thick columns of smoke rising from the surrounding fields.

"The first thing I became aware of," Tuck remembers, "was a smell. A very special smell I hadn't known since the day I was shot down—cordite. It made my pulse leap. I gave a whoop and smacked Zbishek's back, then he smacked mine. There were shells whistling overhead and bursting damned close, but we couldn't have cared less. We felt we were practically home!"

Then came a series of louder, different sounding explosions. They crawled to the end of the loft and looked down into the farmyard. Four German 40 mm.s were parked beneath, close by the wall of the barn, their crews slaving to keep up a rapid fire. From this position Tuck could distinguish in the distance running figures, horsemen, a few lurching vehicles.

It was a beautiful morning! The sky overhead was pale blue, and the early sun was scattering rubies on the snow. Far to the south some sleepy old hills had pulled fleecy coverlets of cloud over their white heads. And those little running figures on the plain to the east were—Russians!

Suddenly the 40 mm.s stopped firing. Powerful trucks swept round into the yard. In a matter of seconds the well-drilled gunners had their pieces hitched on to the trucks and the whole outfit was roaring off down the road to the southwest. Tuck and Zbishek stood up and, unkempt heads sticking out of the wreckage of the barn roof, they waved and cheered like school-children on a holiday.

One of the slave-laborers, a plump and pretty girl named Maria, ran towards the barn shouting: "We are free! The Germans have all gone, and our brothers are here! Come out, come out! We'll give them a welcome together!"

They hurried down, but just as they reached the yard more shells exploded, even closer than before. Maria fled squealing into the house, to join her mother and sister, Shenia, Vasya and several others—already sheltering in the cellar. Tuck and Zbishek dropped into a long, narrow potato trench about three feet deep and lay on their stomachs.

The bombardment lasted for perhaps ten minutes. Bits of tiles and clods of earth fell on their backs now and then. When the din stopped, with an astonishing abruptness that they somehow mistrusted, Tuck rolled on to his back and started to sit up, slowly. He stopped halfway, leaning back on his elbows: he was staring right into the wickedly shining little snout of a Tommy gun.

The weapon was held by a villainous, fur-capped figure—from this low angle he seemed a giant! His light blue eyes were glazed, lifeless. Little rivulets of sweat trickled down his broad, unshaven face. Tuck could see his wide nostrils distending as he inhaled in short, fierce breaths. He was swaying gently from side to side, and the muzzle of the gun weaved about just two feet from Tuck's face.

No doubt about it: this fellow was half-drunk. Tuck had the unsettling feeling that he was already lying in his grave. . . .

Aksinia, don't fail me now!

He forced the stiff muscles of his face into what he hoped was a friendly, confident smile, and shouted in Russian: "It's good to see you, Comrade! We are British officers—escaped from prison camp!" The giant blinked, went on swaying. No change of expression.

Until now Zbishek hadn't realized they had a visitor —he'd raised his head cautiously and peered in the other direction. Now he whirled round and jerked to a sitting position.

Very slowly Tuck started to get to his feet, keeping up his flow of talk. "We owe our lives to some brave Russians, Comrade. They have been looking after us for weeks now, ever since we got away from the Germans. They themselves have been prisoners—slave-workers. Some of them are over there in the house now. I know they, too, will be overjoyed to see you. . . ."

Tuck was out of the trench now, facing the Russian. Zbishek was getting to his feet and about to follow.

From this level the Russian's menacing appearance had diminished. He was no longer a giant. In fact, he was rather squat and potbellied. Over his uniform he wore a very long civilian overcoat of navy blue, and one of his boots had a piece of wire wound round it to keep the sole on. But he was armed like a whole platoon. At his waist: a heavy revolver and an ornamental cavalry saber, so long that its tip trailed in the snow. Slung over one shoulder he had a sack which obviously held grenades. Crossed over his chest were bandoliers of ammunition. And in his big, mittened hands: the Tommy gun. . . .

Making no sudden movements, Tuck undid his coat and pointed to the wings on his faded, grimy tunic. For the first time the Russian's expression changed. He frowned, leaned forward, and peered shortsightedly at the brevet. All the time he kept the muzzle of the gun wavering between the two of them, at stomach level. Very, very slowly a comprehending light crept into those blue eyes, and the ghost of a grin softened that flat face. Tuck took a chance, stuck out his hand and cried: "*Kak pajievietye, Tovarich!*" (How are you living, Comrade?) The Russian slung his Tommy gun over his shoulder and returned the greeting in a gritty bass. Next moment he was throwing his arms round their necks and kissing them full on the lips.

"We are allies!" he rasped. "Together we are smashing the Fascists! I am your friend—Mikhail is my name! And to think I almost shot you!" Then he rummaged in his pockets, brought forth an aluminum flask and thrust it at Tuck.

"*Speert!*"

Tuck took a swig. He had drunk a wonderous variety of alcoholic beverages in his day, including South Carolina "moonshine" and "jungle juice" distilled from cabbages and potatoes in Sagan camp. But this stuff imprisoned the air in his lungs, went down to his boots, bounced back up again, exploded somewhere behind his eyes, scorched the inside of his nose and threw him into a violent fit of sneezing and coughing. Now he knew why Mikhail swayed to and fro like that. . . .

*   *   *

The road was tight with refugees. Most of them bore their shabby possessions on their backs, but some families struggled along with horses and buggies, handcarts, bicycles, sledges or baby carriages. Old men and babes in arms . . . squawking toddlers and bent crones . . . pet dogs and bird cages, picture frames and rocking chairs, cradles and cooking pots, goats and grandfather clocks. The deep ditches on either side were filled with the dregs of battle: upturned carts, dead horses, Russian and German soldiers lying in grotesque, stiff-limbed attitudes. For every dead German there appeared to be two or three Russians: Tuck reflected that the Wehrmacht might be a beaten, retreating army, but they were still fighting efficiently—making the Russians pay a stiff price for their advance. He could look on all these broken bodies without emotion, but the hollow, haunted faces of some of the refugee women evoked a deep, sad anger.

They are the real casualties. . . .

Off the road were mounds of snow and patches of deep mud, so Mikhail, leading them to his company's H.Q., stuck to the narrow verge above the ditch, shouting and shouldering man, woman, child and beast out of his way. He set a surprisingly brisk pace, rolling along on his stumpy legs—still nicely drunk.

After a while Tuck heard the sound of an aircraft, and spotted a Me. 109 doing a lazy turn at about 1,500 feet. He and Zbishek stopped and watched it. From prison camp they'd seen plenty of Allied bombers, but this was the first Messerschmitt Tuck had set eyes on in more than three years—and in an instant he was vibrant, bristling, like an animal scenting its prey. The killer-instinct, still flaring—not any new vengeful urge, no pent-up hatred; just the same old, cold fury, directed not towards man, but machine. . . .

The Messerschmitt dropped its nose and started down towards the road.

"Look out—strafing!" Zbishek yelled.

Wails of terror. The dense throng began to flood over into the ditches. Tuck saw a very old woman hobbling shakily, with no one to help her. He put an arm round her thin shoulders, swept her down the steep slope and shoved her towards the wreckage of a cart.

"Under there, *Babushka!* Quickly!" She fell on her knees and began to crawl under the shattered vehicle. Tuck spotted Zbishek and Mikhail by a burnt-out truck, shouting and beckoning to him. He sprinted over and threw himself flat beside them as the scream of the Messerschmitt's engine rose to a crescendo. Then the clamor of the fighter's cannon made the ground shake, and they were showered with snow and mud.

As the Messerschmitt climbed away, Tuck raised his head and found himself looking straight into the wide eyes of a little boy of about three. The child was sitting erect, against the side of the ditch, but his head

was bent to one side at an odd, cringing angle. Tuck
got up and started to speak to him.

"It's all right, *Brattis,* it's over now. . . ."

Then as he took a step nearer he realized that the
fixed stare was lifeless, and saw the blood oozing from a
hole in the center of the little chest.

"Oh Christ, look at that!" Mikhail shrugged and
pushed out his lower lip. Zbishek put his hand on
Tuck's shoulder. Then they started on again.

Soon the rumble and rattle of carts and the crunching
of weary feet rose once more. Those families who had
suffered damage or casualties pulled off the road with
their possessions to let the others pass. At intervals on
either side little groups shovelled the snow away and
began hacking at the hard-frozen ground to make
shallow graves for their dead.

Another hour, and Mikhail turned off into the fields
crying: "Time we had some food, Comrades! How long
since you had full bellies?"

*How long? Not since the Age of the Great Liz-
ards. . . !*

If you ask Tuck to describe the finest meal he's ever
had, he will tell you about the snack he shared with a
Russian trooper in the lee of a burnt-out tank.

"Within seconds he had a wonderful fire going,
and by the simple process of stuffing snow into a big,
rusty tin he got a couple of pints of water boiling.
Then from somewhere beneath his voluminous coat he
produced something wrapped in old newspapers.
Zbishek and I sat watching, fascinated, while he unwrapped it and slapped on one of his grimy palms a
slab of fat pork which must have weighed about a
pound and a half.

"He hacked the pork into small pieces with a big
jackknife, and dropped them into the boiling water.
Then from another pocket came a large onion, which
he chopped up on the short butt of his Tommy gun
and added to the pot. Lastly, with a broad grin he
conjured up a paper poke full of salt and liberally
seasoned the mixture.

"Zbishek and I were dead quiet, watching the pieces
of pork and onion chase each other around in the

bubbling liquid, inhaling the wonderful aroma. We'd forgotten all about the slaughter we'd witnessed back on the road—our minds were blank, the only thing that mattered in the world was that steaming tin!

"After what seemed an age, Mikhail sharpened a piece of wood and skewered a piece of meat. We watched, our mouths hanging open, as he raised it in front of his face and blew on it several times. I know I was literally drooling as he held the morsel out to me and said: 'Porjowouster, Colonel—it will be as good as your mother's cooking!'

"I took it and popped it in my mouth. The joy of tasting that little bit of greasy pork was quite staggering. Zbishek gave a kind of a groan as he saw the expression on my face, and Mikhail quickly fished out another piece for him. Together we sat chewing, slowly, with our eyes tight closed. As the juice trickled down my throat I felt a warmth stealing right through me and felt I was really starting to live again."

They reached the Company H.Q. about noon—a small cluster of farm buildings and cottages. Outside the biggest building an ambulance unit was at work. About twenty wounded men lay on stretchers under a line of fir trees. Each stretcher rested on a light metal frame with small but wide-treaded wheels, and harnessed in front was a large, heavy-coated dog. Sprawled beside each dog was a soldier wearing a Red Cross armband. Mikhail explained that, with the dog hauling the stretcher, only one man was needed to bring in a wounded soldier. The man crawled beside the dog, and presented a very small target to the enemy.

Some of the "bearers" were, in fact, girls. They were introduced to one of them, Daria—a short and sturdy brunette in a bulky reefer jacket and thick, padded trousers. She seemed to talk and laugh continually. She told them that Mikhail's C.O., a major, had gone to Regimental H.Q., about two miles away, and offered to supply horses and lead them there.

There were no saddles, but it was good to be carried along for a change. As they emerged from some pines and started across a stretch of open country, Daria called: "We had better keep well spaced out from here on. The *Nimietski* (Germans) are only about two kilometers away, and they may take a shot at us."

The *Nimietski* did just that.

Whee-woomf! The horses bucked and then leapt forward as the shells crashed down.

"How I stayed on, I don't know," Tuck says. "Looking ahead I could see Zbishek and Daria charging through the shell-bursts. Zbishek seemed well in control of his mount, but Daria's short legs were flapping madly and she was hanging on to the mane with the reins flying loose. With each explosion her horse would veer to one side and she'd heel over.

"Suddenly one landed very close to her, and she disappeared into the shower of mud and snow. For a moment I thought she'd had it, but as the muck cleared I saw her on her hands and knees, getting to her feet. Her horse was lying a few yards away, kicking at the sky, and blood was spurting on the snow all round it.

"I reined in, and with great difficulty stopped beside her. Just then there was a mad yell and Mikhail thundered past, completely out of control. Daria ran up and jumped across the rump of my horse, landing on her chest and tummy. I grabbed the seat of her pants and yanked her up behind me. Her arms clamped round my waist—and then, damnit, I realized she was howling with laughter!"

He dug in his heels and the horse shot forward again. Ahead lay about a kilometer of open ground. It seemed like a hundred miles. The ground all around them erupted as the shells fell thicker and faster. The terrified horse swerved and half-stumbled. Dirt stung their faces, shell fragments sang past their ears. But when at last they reached the shelter of a group of farm buildings and managed to halt their mounts, Daria was still laughing. . . .

Zbishek stood in a doorway, yelling at them above the din of the bombardment.

"Down here—the cellar!"

As Tuck pushed the girl ahead of him through the doorway there was a tremendous explosion behind him. He was lifted off his feet and thrown down a flight of steps into a crush of bodies. When they'd disentangled themselves Daria stopped laughing long enough to wring his hand and say: "Colonel, you are a fine horseman! We must go riding together again some time. . . ."

*   *   *

Up until now Tuck had got along splendidly with the Russians. Shenia, Vasya, Maria and the other slave-laborers had kept them alive in the midst of the enemy; Mikhail had turned out a decent enough chap, and this madcap Daria was vastly entertaining. It looked as if all his ideas about the Russian nation were going to be fulfilled. They were tough, they were highly courageous and, it seemed, extremely friendly. Even Zbishek seemed to be warming to them.

The change, the rude awakening, came when the German bombardment ended and Mikhail led them to his major. The moment they entered the roomful of officers they could sense suspicion and enmity.

"It was a large room, and there were about ten officers. They all wore bright, flashy epaulets and had ancient sabers buckled at their waists. At one end, behind a table which was bare except for a field telephone, stood the major.

"None of them moved or spoke as we walked the length of the room up to the table. I suddenly realized that I felt much less at my ease here than I had when I'd been shot down in France and marched before the officers of the German Flakregiment!

"I looked at some of the faces. Tough as nails, but brutish—horribly unintelligent. Zbishek is nearly six feet two inches tall, and I am just about six feet; all of these men were either pretty short or medium-sized. It is possible that, in trying to overcome our uneasiness, and despite our unkempt appearance, we might have given an impression of arrogance."

Mikhail saluted and cleared his throat.

"These are the two Englishmen," he said.

The major, a slight man with deep-set eyes and a lean, bone-white face, looked straight at Tuck and shouted at the top of his voice: "I don't care who the devil they are, I will not have people making a noise and getting excited in my headquarters!"

So far neither of the R.A.F. men had spoken a word, and certainly none of the others in the room was making a noise or showing any sign of excitement!

Tuck looked the major straight in the eye and said slowly, firmly, in Russian: "How extraordinary!"

They stood very still and upright, staring at each other. There wasn't a sound in the room except a sudden squeak from Mikhail's poor old boots as he shifted his weight from one foot to another. Suddenly the major closed his eyes and with a sigh slumped wearily into his chair.

"Forgive me," he grunted. "I have not slept since I don't know when, and I feel rather ill."

"Please don't worry, major," Tuck said. "We understand."

"My officers and I have the honor of serving in the Second Gvardia Regiment, and this, as you probably know, is the First Ukrainian Army Front, commanded by the great Marshal Koniev. Now we shall require some information from *you!*"

Tuck introduced himself and his companion and outlined the story of their escape. At the name "Captain Kustrzynski" all eyes swung on to Zbishek. Tuck, anticipating awkward questions here, managed to slip into his account the story that though the Captain had Polish blood, he was in fact in every respect British. But he had the sinking feeling that none of them were convinced.

"Our Intelligence men will wish to talk to you later," the major said. "In the meantime we will find you somewhere to sleep."

From then on they were not exactly under guard, but they soon became aware that their every move was being watched and that an armed soldier was always hanging around in the background. They stayed with the Second Gvardia Regiment for two weeks, advancing with them—going back west, instead of east. Their repeated requests for a message to be sent to the British Air Attaché in Moscow were ignored, and they were allowed no transport.

They were, virtually, prisoners again.

"Right, then," said Tuck, "we'll bloody well escape again!" But that was very difficult—they were never left alone.

So, against their will, they fought for Stalin—taking part in some very stiff battles against die hard S.S. units, flushing snipers out of woods and farms, enduring bombardments and strafings, pushing the *Nimietski* back a few miles every day. It was galling to have to fight to gain every few hundred yards in the wrong direction!

The Russians suffered appalling casualties, but it didn't seem to bother them. Russia was rich in men—and women. (It was only after two or three days that Tuck and Zbishek realized that some of the soldiers that fought with them were females: they dressed exactly the same as the men.)

They were switched from company to company—always accompanied by at least one officer and several soldiers—and between actions were frequently entertained at huge feasts. Whenever their hosts got hungry they simply sent out a party to round up a few cattle, then roasted about twice as much meat as they needed. There appeared to be no shortage of vodka or wines. Sometimes the drinking lasted for twenty-four hours or more.

Once they were taken to the First Ukrainian Army H.Q. to be interviewed by General Alexandrov, one of Koniev's commanders. Their hopes soared—surely the General would arrange for them to be "cleared" and sent on their long journey home without further delay? But when they got there they found the Headquarters officers all hopelessly drunk—including Alexandrov. They were brought back, protesting, to the front.

After a two-day battle against an S.S. unit, one survivor was captured—a boy of about eighteen who, the Russians somehow discovered, was a fine pianist. They dragged this lad into a farmhouse taken over as the officers' mess, dumped him in front of an ancient piano and told him: "As long as you keep playing, you live. As soon as you stop we will kill you!"

The youth was already exhausted by days of hard fighting, but he played for sixteen hours without respite. Chopin, Liszt, Grieg, and an assortment of light music and popular songs. He really was an excellent pianist. His captors flocked round the piano, humming and singing. They clapped him on the back, fed him drinks until he was half-tipsy. And when finally he slumped forward over the keyboard, sobbing and moaning, they picked him up, carried him outside, propped him against the back of the house and blasted him with Tommy guns.

Zbishek had to argue hotly with Tuck to prevent him flying into a rage over this brutality.

"For God's sake, Robert, nobody in England knows yet that we've escaped. If you insult these bloody barbarians they're liable to give us the same treatment—and who'd ever know we weren't shot by the Huns?"

They were taken to the First Ukrainian Army H.Q. again, this time to be interrogated by Intelligence officers. It was a long and unpleasant session. The interrogators were obviously convinced that Zbishek was a Polish national who'd escaped to England and joined the R.A.F.—which was, of course, the truth. They made it plain that the Soviet government intended to "look after" the Poles who came within its area of authority, and that none would be allowed to return to Britain or any of the Western countries.

They also asked a great many technical questions about the latest British and American aircraft, but since both the R.A.F. men had been out of the war for years this was rather pointless. Some nasty phrases crept in —"imperialist agents . . . capitalist leaders . . . corrupt governments of the West. . . ."

Could this be the glorious ally? Was this the kind of doctrine that had brought these armies a thousand miles from Stalingrad? Where was the old, mystic Russia of Aksinia's tales? Tuck was sick with disappointment.

Zbishek was drawn, depressed. Despite all his care and vigilance—only once, in the heat of battle, had he lapsed into Polish—it looked as though he'd failed. He was convinced now that Russia would more or less annex Poland. He couldn't tolerate the idea of being sent to live under a Communist puppet government. He had a dream about going to start a new life in Canada or the U.S.A.

"We've got to get away—at once," Tuck said, "even if it means taking a big chance." The sad Pole rubbed his drawn cheeks and spoke softly, thoughtfully.

"If we're caught they'll call us 'deserters'—or they'll say that by running away we've proved that we're spies. As soon as we're missed they'll put their bloody N.K.V.D. on to us."

"If only we can get word to the Air Attaché in Moscow, then we'll be all right. What d'you say, Zbishek old boy? Shall we have a try?"

Zbishek sighed and regarded Tuck with dull, unblinking eyes.

"Well, here we go again," he said.

*   *   *

They got away by climbing out of a window during a thick blizzard. For the first night and the whole of the following day they slugged eastwards through the swirling snow, and Zbishek's dysentery got worse.

They had no papers, and the Russians had checkpoints on all the railway stations and main roads. Zbishek was in his homeland now, but it had so changed that he was not sure that he could trust any of his countrymen. Poland had become a Communist state. They steered clear of towns so far as possible, and asked help from no one.

They believed that if only they could get over the Soviet frontier it would be possible to get through to the British Embassy in Moscow by telephone without delay or form-filling. Once they'd made contact with the Embassy they were safe.

Sometimes they were able to hitch rides on goods trains, and even on solitary Soviet army trucks. Zbishek grew still more depressed, but Tuck forced him along. In those final days the Pole was wholly impelled by Tuck's vigor and determination.

Always Tuck had shone brightest in the toughest situations. He was once more at his best now.

In the prison camp he'd been overshadowed by the greatness of Roger Bushell. He was an outstanding leader only when there was plenty of action; he was no good at politics, administration or long-term planning. But when it came to flying or fighting Tuck, like Bader, was supreme.

And he never fought a harder fight than on that last, long trek through the Polish winter—because by now even his iron health was beginning to wilt. But as ever, he was saved in the end by that mysterious ally, "Tuck's Luck." This time it came up with what is perhaps the happiest, most perfectly-timed coincidence of his career.

A chance in many millions——

They had to pass through the town of Czenstowskova. In the streets they realized they were being shadowed by a small, rat-faced man in a black leather coat and an enormous fur cap. They tried to shake him off by diving into the thickest crowds, and then doubling back through alleys, but it didn't work.

Suddenly the little man quickened his pace, drew abreast and fell into step with them. Tuck expected to feel a pistol jab into his ribs and hear that dreaded word: "Papers!" What he did hear sent him reeling.

"Wotcher, Guv!" said the little man. "Stanford Tuck, ain't it?"

Tuck staggered to a wall, leaned against it and took several deep breaths, then turned and regarded the stranger.

"Who the hell are you?"

"Dickinson's the name, Joe Dickinson. Don't suppose you've ever 'eard of me, but I was in 'the bag' along of your bruvver, Jack. We was nabbed together at Calais, matterafack."

"But . . . but how did you——?"

"S'easy, Guv—Jack 'ad a picture of you, kept it by 'is bunk. Talked 'bout you all the time, old Jack did."

Tuck ran a hand through his thick, matted beard and glanced down at his filthy, tattered coat. He would have sworn that his own mother wouldn't have known him right now!

Zbishek began to laugh, high and shaky. "I don't believe this!" he gasped. "It's a mirage!"

There were people pushing past them, and here they were talking loudly in English. Russian uniforms and the new Polish Communist police mingled with the crowds. It was all like a daft dream. The little man lit two cigarettes and handed one to each of them. Tuck took a long drag. It only made his head spin more.

"Guv, I think you and your mate oughta come along 'ome with me. Pair of you look in a bad way."

"H-home?"

"S'right." He smiled broadly, linked arms with them and started on down the crowded street. "I've got it all taped 'ere, gents. Been 'ere—oh, must be two years now, ever since I got outer the cage, see? Got me a big fat Polish girl, some cows and chickens. I 'ope this bleedin' war *never* ends!"

"What did you say your name was?" Tuck asked.

"Dickinson."

"Mr. Dickinson, I'm very pleased to meet you."

They stayed several days at Dickinson's little farm, and Mrs. Dickinson (Tuck called her that, though he had serious doubts about the legality of their union) fed them on steaks and chicken, home-made bread and cakes. They put on a little weight.

"See wot I mean?" Dickinson said every time a meal was served. "Be no bloody victory for me if I 'ave to give all this up and go back to flippin' Poplar!* Never live in a city again, not me. So when you get back, gents, mum's the word—I'm dead, see? I do you a favor, you do me one, eh? And lemme give you a tip—never mind them West End tarts, find yourself a girl that 'as a farm. Look to your old age, I always say. Man stays 'ungry for food a lot longer than for the other thing!" (As far as anyone knows, Dickinson is still there today, though his name appears on his regiment's roll of honor and on a London war memorial.)

---

*A slum section of London near Whitechapel.

Dickinson even organized papers and railway tickets to the frontier, and supplied them with fresh clothes and money. And then, a couple of nights before they were due to leave, he brought to the house a girl whom he introduced to them simply as "Addie."

Addie was very beautiful: high cheekbones, large and slightly slanted eyes, yellow hair drawn tightly back to the nape of her long neck. Addie was all-woman, sweet-moving and maturely poised: it was hard to believe that she'd been a member of a Catholic partisan group which had been practically wiped out at the time of the abortive Warsaw uprisings, and that she was now on the run from the Communists.

Addie was really Mrs. Adrek. Her husband was a sergeant in a Polish squadron in Britain.

She sat down by the fire, folded her hands in her lap and said, "Colonel, Captain, will you take me with you?"

"What—to England?" Zbishek croaked.

"Yes. My husband is there. Where else would I want to go now?"

"Sorry, it's . . . out of the question," Tuck said.

"But I've worked out a plan, Colonel. Will you not hear it?"

They listened. It was a good plan. After hours of discussion they finally agreed to it, though it certainly reduced their own chance of success.

Afterwards Tuck was confused and angry—he hadn't meant to agree under any circumstances. It was ridiculous to involve themselves with a wanted partisan at this late stage when they had only one short jump to make to reach the frontier. The girl's presence would complicate everything and could provide the Russian police with strong grounds for arresting them even after they'd contacted the Embassy. . . .

It dawned on him that he found Addie extremely attractive—and that made him even more angry with himself.

* * *

Tuck and Zbishek, shaved and in their fresh clothes, traveled on the train as workers. Fortunately the train was packed and there was no time to check papers. They crossed the frontier at Slupia, and got off at a small town just inside Soviet Russia. After all the worry, it had been amazingly simple. . . !

Tuck went straight to a public telephone, and after a great amount of tedious argument, and a bit of blustering too, managed to get a call through to the office of Wing Commander Shaw of the British Air Mission to Moscow. Shaw, when he got over his astonishment, boomed delightedly, and it was a wonder just to listen to his English voice.

He was a pilot, and Tuck had met him once or twice, casually, early in the war—in the "men only" bar of "the Airmen's Boozer" (the magnificent, high-vaulted Royal Air Force Club in Piccadilly), and at Hornchurch. Tuck remembered him as a neat, brisk man with a deep, warm laugh.

But Shaw wasn't laughing now——

"My God, old man, we heard you'd been shot!"

"Who told you that?"

"Jungle telegraph—Underground boys were sure you'd had it."

"*Well* . . . I take pleasure in issuing an official denial!"

Shaw chuckled. "No need—these were only hearsay reports, they weren't given out to the Press, or next of kin. Now then, can you make your own way here —to Moscow?"

"Yes, I'm sure we can."

"Fine. Meanwhile I'll wire London. I'll have to know some of the facts, though. Let's have the story, briefly for now. . . ."

Tuck outlined their experiences since their escape from the barn at Bransdorf. He talked rapidly, disjointedly; relief was an exquisite excitement that dulled his mind and loosened his tongue, like a powerful cocktail on an empty stomach. But he was careful to make no mention of Addie—waiting for them at Czenstowskova, trusting them to keep their promise. . . .

When he hung up a good twenty minutes later, he

and Zbishek knew they were free men once more—
provided their daring plan to pick up Addie and smuggle her to England was not discovered. He'd said nothing about her to Shaw, because he knew no diplomatic
representative could be party to any such scheme.

They traveled openly on the train to Moscow, without trouble. Wing Commander Shaw, his colleagues of
the Air Mission and Embassy officials quickly established the identity of both officers, arranged sea pas-
sages home from Odessa, and helped them send mes-
sages to their kinsfolk and friends.

During their stay in the Soviet capital they attended
many memorable parties and met quite a few top
Russian pilots and one or two political bosses. Tuck
was angered to discover that most of them knew nothing about Dunkirk or the Battle of Britain, and the
others regarded Britain's ordeal in 1940 as quite unimportant. For them the war began in June 1941, and
victory—at Stalingrad. They had a hopelessly unbal-
anced conception of the war's history. Tuck was glad
when it was time to leave.

They went back by way of Czenstowskova and picked
up Addie. She was dressed as a boy. Tuck grieved for
the long golden hair she'd cut off, wondered how on
earth she'd managed to flatten her high and prominent
bust. She stayed with them on the long, slow journey to
Odessa, and though she had no papers she was never
questioned. And she was careful to keep out of sight
whenever police or an Embassy official came near.

But her pose had to end at the Odessa docks, because her name wasn't on the list of hundreds of refugees and former prisoners scheduled to embark on the
liner *Duchess of Richmond* for Southampton. And
there were Russian sentries and customs officials at the
bottom of the gangway, British guards and ship's officers at the top.

They decided to watch the docks that evening, to
establish how well the gangplank was guarded and
whether there might be a way to get her on board during the night.

Zbishek stayed in his room to rest and have a

Russian steambath. Tuck and Addie went down to the waterfront and found a shadowy place under a crane platform from which they could watch the *Richmond*. She was strongly guarded. They waited, hour after hour, hoping that late in the evening the number of guards might be reduced.

And as they waited they talked. Tuck, in spite of himself, had become more and more fond of this remarkable young woman. He had found that he could talk freely with her—and there was only one other woman with whom that had been possible. Come to think of it, in many ways she was like Joyce—cool, strong, patient. . . .

He never knew how it came about, it was without an instant's forethought. Suddenly she was in his arms, clinging, warm, trembling; only for a few seconds, then she put her hands on his chest and said quietly—*coolly*—"No, Robert . . . it's no good." He let her go, and she moved away to the edge of their hiding place. The pale moonlight caught her. He saw that her face, so small under the cropped hair, had an expression he hadn't seen before. Tired, tired. He felt like that himself. He walked up behind her and put his hands on her shoulders. She reached up and placed her right hand over his, holding it there, pressing it down. He could feel her still trembling.

"It's no good," she said, "and you know it. Please never do that again. It hurts, Robert, it damned well hurts!" He knew very well he had only to turn her around and pull her to him once again, and they'd both be lost. But he didn't.

She had a husband to find again, he had Joyce.

"Let's get back," he said. "It's getting cold."

They decided on a plan to smuggle her on board in broad daylight. In an Odessa market they bought a lot of fruit, packets of biscuits, bottles of wine and all sorts of cheap and silly souvenirs—the bulkiest they could find. When they walked to the gangway at 11 a.m. on the 26th March, 1945, she followed, carrying the parcels. The Russian customs men ignored the parcels, and their sentries let her through, but the British were sticky. Nobody could come on board without an embarkation card.

Tuck argued for five minutes, making himself thoroughly objectionable, before they gave way. Within a quarter-of-an-hour they had her stowed deep in the darkness of the forward hold, along with the food and wine. Soon after 11:30 the British guards and officials were relieved by another shift. Nobody noticed that the "boy porter" hadn't returned ashore.

They sailed next day, and when they were well out to sea and had dropped the pilot, Tuck went to the captain. He knew that the ship would be searched for stowaways, and that if Addie were discovered she'd be put ashore at one of their ports of call and sent back on the first boat. He was going to gamble everything on a frank appeal, hoping that the captain might make an exception or find some loophole in the regulations.

Another stroke of his fantastic luck: the master of the *Duchess of Richmond* turned out to be an old and close friend of one of Tuck's cousins—Donald Dunn, a distinguished sea captain who'd served many years in the same line. . . .

"The ship will have to be searched," he said, "that's essential. But the regulations don't say *where* the search should take place. So if I leave it until we've passed Gibraltar, then the young lady can only be put off at one place—Southampton. Meanwhile you'll have to look after her, take her hot meals. Now, if you go to the galley and ask for . . ."

And so they came home.

Zbishek Kustrzynski—to become a British subject, and eventually to emigrate to Canada, where he found a congenial job at a textile factory in the new "boom"-town of Cornwall, Ontario.

Addie—to find her husband within a few days, and in time to become a happy and efficient housewife.

Bob Tuck—to Joyce. They were married by special license within a week.

She was waiting for him when he landed, smiling a small, very intimate smile. A tall, strong, straight-backed girl with agate eyes. They didn't speak until they'd got away from the reporters and the R.A.F. reception officers and she was driving him off in that same little Morris. Then——

"Where are we going?" he asked.

"Nowhere, Robert. Nowhere, for a long while."

ABOUT THE AUTHOR

Born in Glasgow, Scotland, in 1924, LARRY FORRESTER dropped out of school at fourteen to work as a training reporter for a local newspaper during the day, while playing drums with a jazz band at night. With the outbreak of World War II, he volunteered for flying duties with the Royal Air Force and, under the Admiral Towers Training Scheme, was sent to the United States and became a U.S. Navy flight student, winning his wings at Pensacola, Florida. He flew operationally in Europe and later in Southeast Asia. Demobilized in 1946, he worked first as a reporter and feature writer for Express Newspapers, London, then as a sub-editor for Reuters Wire Service. But flying was still in his blood, and in his spare time he wrote *Fly for Your Life*, which became a European bestseller and which has been constantly reprinted since it was first published in 1956. Mr. Forrester's other writings include the screenplay for the Pearl Harbor epic *Tora! Tora! Tora!*, and more recently *Rehearsal for Armageddon*, a two-hour special for NBC. He is currently under contract with M.C.A., Universal Studios.

Join the Allies on the Road to V...
BANTAM WAR BOOKS